INTERNATIONAL SERIES OF MONOGRAPHS IN
AERONAUTICS AND ASTRONAUTICS

DIVISION III. PROPULSION SYSTEMS INCLUDING FUELS

VOLUME 6

PRINCIPLES OF SPACEFLIGHT PROPULSION

PRINCIPLES OF SPACEFLIGHT PROPULSION

by

E. M. GOODGER
M.Sc. (Eng.), Ph.D., M.I.Mech.E., A.M.I.E., Aust., A.F.R.Ae.S., F. Inst. Pet.
Sometime Professor of Mechanical Engineering, The University of Newcastle, N.S.W.

PERGAMON PRESS
OXFORD · LONDON · EDINBURGH · NEW YORK
TORONTO · SYDNEY · PARIS · BRAUNSCHWEIG

Pergamon Press Ltd., Headington Hill Hall, Oxford
4 & 5 Fitzroy Square, London W.1
Pergamon Press (Scotland) Ltd., 2 & 3 Teviot Place, Edinburgh 1
Pergamon Press Inc., Maxwell House, Fairview Park, Elmsford, New York 10523
Pergamon of Canada Ltd., 207 Queen's Quay West, Toronto 1
Pergamon Press (Aust.) Pty. Ltd., 19a Boundary Street,
Rushcutters Bay, N.S.W. 2011, Australia
Pergamon Press S.A.R.L., 24 rue des Écoles, Paris 5e
Vieweg & Sohn GmbH, Burgplatz 1, Braunschweig

First edition 1970

Library of Congress Catalog Card No. 77-88306

Printed in Hungary

08 013884 5

TO MY FATHER
WHOSE ENCOURAGEMENT AND SUPPORT
WERE UNFAILING

". . . the basic spaceflight problem . . .
is one of propulsion . . ."
A. V. CLEAVER

CONTENTS

PREFACE

THE aim of this book is to show how one of the major barriers to space exploration is being surmounted. The achievement of very high velocities of escape order is the essence of spaceflight, and adequate propulsive thrust constitutes the prime requirement. Satellites and spaceprobes have already proved much of the feasibility of manned spaceflight, whereas lunar landings and interplanetary voyages are predicted within the very near future. Interstellar missions may then be expected as a natural extrapolation.

Spaceflight propulsion is now a sufficiently established and respectable technology to have engendered its own library of textbooks, many of which adopt a fairly advanced level of scientific knowledge as a starting-point, and/or include such topics as the history of rocket flight, design features of propulsive and associated systems, and detailed examples of rocket hardware. The present volume is intended as a primer, confining itself to the fundamentals of those processes of energy release and conversion which lend themselves to flight within the near vacuum of space. It illustrates the state of the art in this aspect of space technology by providing a general picture of the propulsion techniques available and conceptual, and by comparing their effectiveness and probable fields of application. A review of this kind may well serve as a preliminary step for the engineering student, and for the practising engineer with responsibilities in other branches of the profession, who have an interest and desire to keep abreast with developments in this exciting field of endeavour.

I am indebted to former colleagues of the College of Aeronautics, Cranfield, and the University of Newcastle, N.S.W., for much valued assistance, particularly to Mr. J. G. Alva of Newcastle, and to Mr. D. S. Carton, now of the Aston College of Technology, Birmingham for countless discussions, interrupted on occasions by distance and time. I am grateful for approval to draw upon material published by the Institute of Fuel and the Institute of Petroleum on my behalf. Thanks are due also to my family for their forbearance, and to Mrs. Margaret Birch of Melbourne for her typing skill.

Woburn Sands E. M. GOODGER

UNITS

In any technical work, the selection of units is expected to meet the requirements of consistency and current usage. In the present work, this is particularly important in view of the several disciplines encountered, with dimensions ranging from the nuclear to the astronomic, and also the growing adoption of the International System (SI), with the basic units of metre, kilogram, second, ampere, degree Kelvin and candela (m, kg, s, A, °K, cd) only. In this system, the derived units are coherent in that the product or quotient of any two unit quantities is the unit of the resultant quantity. Derived units of interest here are those for force (newton, N), energy (joule, J), power (watt, W), electric charge (coulomb, C), and electric potential (volt, V). This system differs from the metric technical system which incorporates additional basic units.

Metric units are already in general use in dealing with the statics and dynamics of space, and with quantities involved in electrical propulsion, hence conversion is rarely required in these cases. In conventional thermal rocketry, however, the aeronautical units of foot-slug-second (ft-slug-s) still apply. Furthermore, the pound mass (lbm) from the British Absolute ft-lbm-s system is still used, incorporated in the mixed ft-lbf-lbm-s system, for the parameter "specific impulse", together with the Centigrade heat unit (CHU) for energy. The centimetre-gram-second (cm-g-s) system is commonly used on the atomic scale, and the energy unit of kilocalorie per mole (kcal/mole) in thermochemistry. In these special cases, therefore, the current conventional units have generally been employed, with conversion to SI where necessary, as in the following Tables.

Unit Conversions

Quantity	ft-slug-s	ft-lbm-s	SI equiv.	cm-g-s equiv.
Length	1 ft	1 ft	0·3048 m	30·48 cm
Mass	1 slug	32·174 lbm	14·594 kg	14,594 g
	0·03108 slug	1 lbm	0·4536 kg	435·6 g
Force	1 lbf	32·174 pdl	4·4482 N	$4{\cdot}4482\times10^5$ dyne
Energy	1 ft.lbf	–	1·356 J	$1{\cdot}356\times10^7$ erg
Power	1 ft.lbf/s	–	1.356 W	$1{\cdot}356\times10^7$ erg/s
	1 hp	–	745·7 W	$7{\cdot}457\times10^7$ erg/s
Electric Charge	1 coulomb (C)	1 C	1 C	3×10^9 e.s.u., or statcoulomb

Length equivalents

1 Angstrom = 1 Å = 10^{-10} m
1 micron = 1 μ = 0·00039 in. = 10^{-6} m
1 in. = 0·0254 m
1 statute mile = 1·6093 km = 1609·3 m
1 nautical mile = 6080 ft = 1·1516 statute mile = 1853·6 m
1 astronomical unit = 1 a.u. = 93×10^6 statute mile = $149{\cdot}7\times10^9$ m

Physical constants

Radius of Earth = 3960 statute mile = $6{\cdot}4\times10^6$ m
Radius of Moon = 1080 statute mile = $1{\cdot}7\times10^6$ m
Earth–Moon distance = 238,000 statute mile = 384×10^6 m
Velocity of sound in air at 0° C = 1087 ft/s = 331·3 m/s
Velocity of light in vacuum = c = $9{\cdot}84\times10^8$ ft/s
= 186,330 mile/s = $2{\cdot}9979\times10^8$ m/s
Gravitational constant = g = 32·174 ft/s² = 9·81 m/s²
Electron rest mass = m = $9{\cdot}1086\times10^{-31}$ kg
Atomic mass unit = amu = $1{\cdot}6603\times10^{-27}$ kg
Electron charge = e = $1{\cdot}6021\times10^{-19}$ C = $4{\cdot}8028\times10^{-10}$ esu
Universal gas constant = G = 2781·6 ft-lbf/lbm-mole, °K
= 8310 J/kg-mole, °K

Energy Conversions

	kW.h	hp.h	kcal	CHU	ft.lbf	joule*	MeV
1 kW.h	1·00	1·3405	860·0	1896	$2{\cdot}654\times10^{6}$	$3{\cdot}6\times10^{6}$	$2{\cdot}25\times10^{19}$
1 hp.h	0·746	1·00	641·56	1414	$1{\cdot}98\times10^{6}$	$2{\cdot}69\times10^{6}$	$1{\cdot}68\times10^{19}$
1 kcal	$1{\cdot}163\times10^{-3}$	$1{\cdot}559\times10^{-3}$	1·00	2·2046	3087	4186·8	$2{\cdot}61\times10^{16}$
1 CHU	$5{\cdot}274\times10^{-4}$	$7{\cdot}07\times10^{-4}$	0·4536	1·00	1400·4	1898·4	$1{\cdot}185\times10^{16}$
1 ft.lbf	$3{\cdot}768\times10^{-7}$	$5{\cdot}051\times10^{-7}$	$3{\cdot}239\times10^{-4}$	$7{\cdot}141\times10^{-4}$	1·00	1·356	$8{\cdot}47\times10^{12}$
1 joule*	$2{\cdot}78\times10^{-7}$	$3{\cdot}72\times10^{-7}$	$2{\cdot}39\times10^{-4}$	$5{\cdot}268\times10^{-4}$	0·7375	1·00	$6{\cdot}25\times10^{12}$
1 MeV	$4{\cdot}45\times10^{-20}$	$5{\cdot}96\times10^{-20}$	$3{\cdot}83\times10^{-17}$	$8{\cdot}44\times10^{-17}$	$1{\cdot}181\times10^{-13}$	$1{\cdot}6004\times10^{-13}$	1·00

* SI unit. 1 joule = 10^{7} erg.

Specific impulse:	1 lbf.s/lbm	= 9·8067 N.s/kg
		≡ 32·174 ft/s = 9·8067 m/s
Pressure:	1 lbf/in^2	= 6894·76 N/m^2
Velocity:	1 mile/h	= 0·447 m/s
Specific energy:	1 CHU/lbm	= 4186·8 J/kg
Power:	1 hp	= 745·7 W
Weight:	1 tonf	= 9964 N

1 barn = 10^{-28} m^2
Dielectric constant of vacuum = $\varepsilon_0 = 8{\cdot}854\times10^{-12}$ F/m
Avogadro's number = $N_0 = 6{\cdot}0247\times10^{26}$ molecule/kg-mole
Planck's constant = $h = 6{\cdot}625\times10^{-27}$ erg-s = $6{\cdot}625\times10^{-34}$ J-s
Rydberg constant = R = 109,600
Boltzmann's constant = $k_B = 1{\cdot}3804\times10^{-23}$ J/atom, °K

CHAPTER 1

FUNDAMENTALS

1.1. Introduction

Energy is being used by mankind in increasing quantities and for an increasing variety of purposes. Energy sources have been discovered and exploited since earliest times, and the story of this progress represents a vital part in the development of man himself. In addition to meeting the basic needs of warmth, cooking and illumination, there is evidence of prehistoric use of energy for the enrichment of soil, the smelting of metals, and the manufacture of crude forms of building brick, cement, pottery and coloured glass.

Curiously, one of the earliest applications of inanimate energy to the transport of material was the use of rockets for military purposes in the thirteenth century, since this form of propulsion is also considered one of the latest systems of its kind. However, the emergence of the steam locomotive at the beginning of the nineteenth century heralded a new and particularly significant application of energy—mechanical self-propulsion with man-carrying capabilities. Development in this sphere has shown a marked exponential rise characteristic of most technologies, and has culminated in the era of space travel.

With no ambient medium to provide a reaction force, most systems for propulsion in space are based on the reaction thrust from an ejected jet, although explosion pressure waves and action-at-a-distance forces are envisaged in certain other concepts. In the field of materials used for jet propulsion, an important distinction should be drawn between the energy-storage material and the propulsive working fluid, i.e. between reactant and propellant. At the time of writing, the chemical rocket is the only propulsive device developed to operational standard, and the material used for propulsion combines the two functions. The former function is effected by fuel and oxidant materials stored (usually separately) in their liquid phases or, alternatively, combined in a solid phase, and the latter function by the gaseous products of chemical reaction. The indiscriminate use of such terms as fuels, oxidants and propellants, therefore, raises no serious practical problems to date but, in many alternative systems under development, different materials are envisaged for the two functions. In this book, therefore, these two materials are described respectively as reactants and propellants.

The only two types of reactant likely to be transportable within the space vehicle appear to be the chemical and the nuclear. On the other hand, propellants include chemical reaction products, heated gases, plasmas, ions, electrons and photons. Such a separation of materials brings greater flexibility and improved performance since less compromise is needed in selection for required properties. However, complexity and weight tend to be increased.

A co-requisite for all self-propelled vehicles is adequate service energy for the operation of such auxiliary and ancillary equipment as instruments, controls and telemetry. Details of direct conversion of electrical energy for such purposes in space flight are outside the field of this book, but a brief comment is included as an introduction to further reading, and since continued development may realize an energy level sufficient for limited propulsive purposes.

An interesting feature of a study of this nature lies in the dimensional scales involved since these range from the nuclear to the astronomic. Although many physical laws apply throughout, others (e.g. relationships between particles and waves, position and momentum, space and time) are confined to their dimensional régime. Particular care is necessary with the systems of units employed, and a note on these is given as an introductory section.

After a brief survey of the fundamentals involved in jet propulsion in a space field, the processes of energy release are examined from the bases of atomic and nuclear structure. Subsequent chapters deal separately with propulsion systems based on chemical and nuclear energy release, and on the intermediate conversion of nuclear to electrical energy. Some additional concepts are outlined briefly, and the final discussion includes a survey of propulsion systems.

1.2. The Environment of Interplanetary Space

Our solar system with the Sun as its central star forms a minute component of the Milky Way galaxy comprising many millions of celestial bodies. Earth-launched vehicles designed to explore neighbouring planets will need to commence their operational life at the relatively dense atmospheric conditions near sea-level, and to traverse vast distances of ultra-high vacuum. Although atmospheric density will attenuate extensively with increasing altitude, the vacuum in deep space is not expected to be absolute. On the contrary, the increasing amount of data from recent space probing indicates the presence of particulate matter at very large distances from the Earth's surface. In fact, some conceptual schemes of space exploration are based upon the collection and utilization of this material.

Space vehicles, their propulsive systems, reactants and propellants will all be influenced to some extent by the environment in which they work. This environment is characterized by the physical properties of the medium, the electromagnetic radiation fluxes, and force fields from a distance.

The most significant properties of the medium are density, pressure and temperature. Measurements of these parameters have been collated and averaged to provide a representative model for purposes of calculation and project design. The U.S. Standard Atmosphere(5) represents a model of the lower regions, whereas the ARDS and Smithsonian models attempt to extend these data to extreme altitudes. Figure 1.1 is a composite presentation of these data to indicate the ranges of atmospheric density, pressure and temperature to which the vehicle initially will be subjected.

Atmospheric density falls from its sea-level value of 0·0766 lbm/ft^3 (1·228 kg/m^3) through a value less than 6 per cent of this at an altitude of 25 km, and the vehicle may be considered to be "in space" and free of aerodynamic drag and kinetic heating effects at an altitude in excess of about 200 km.

Atmospheric pressure falls in a comparable manner from its sea-level value of 14·7 lbf/in² (101,400 N/m²) through about 0·5 lbf/in² at 25 km. As shown in the next section, the thrust given by a rocket-type jet

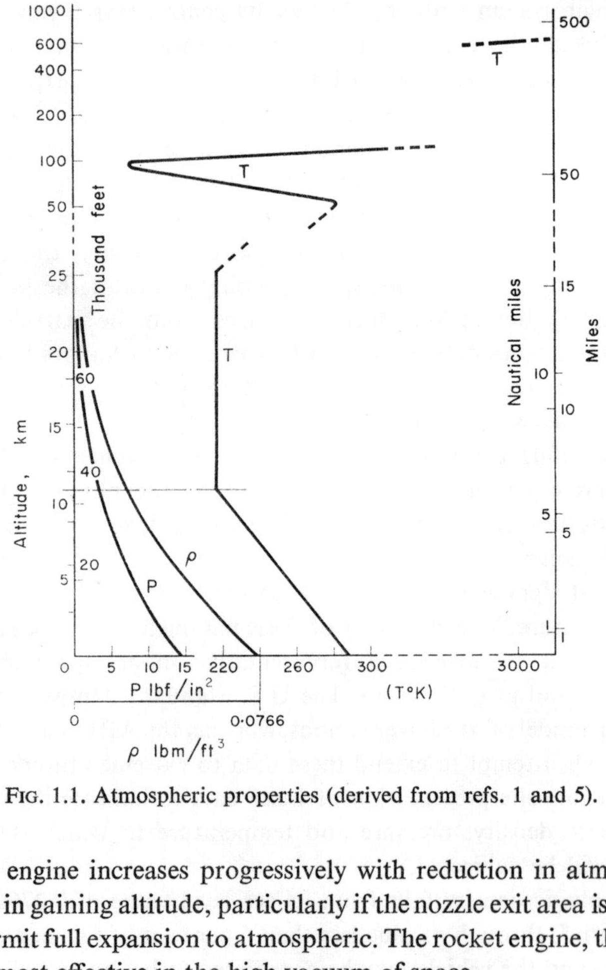

FIG. 1.1. Atmospheric properties (derived from refs. 1 and 5).

reaction engine increases progressively with reduction in atmospheric pressure in gaining altitude, particularly if the nozzle exit area is increased to permit full expansion to atmospheric. The rocket engine, therefore, is at its most effective in the high vacuum of space.

After the initial fall in atmospheric temperature from its sea-level value of 288° K to 216·5° K at the tropopause (11 km), a constant value is considered to hold up to about 25 km in the stratosphere. A steady rise then occurs to about 283° K at 50 km followed by a fall to about 170° K at 90 km, and a probable rise through about 3000° K at 600 km. At these high altitudes, of course, the atmosphere becomes very attenuated, and the high temperatures apply to discrete molecules widely separated from each other. This situation is represented by the term "molecular scale temperature". Interstellar space is estimated to have a black-body temperature of only a few degrees above absolute zero. The equilibrium thermal conditions of a spacecraft will be limited by the incident and emitted radiations, and by the generation of heat within the craft itself. Under conditions of high vacuum, radiation is the only significant mode of heat transfer between the vehicle and its environment.

In manned spaceflight, ranges will be limited initially to the nearest planets within the solar system, consequently the main source of incident radiation to which spacecraft will be exposed will be the spectrum of electromagnetic radiation resulting from the fusion of helium in the Sun. Proximity with planets will add the albedo, which is the fraction of incident solar energy reflected by the planet, together with heat emitted due to the temperature of the planet surface. This temperature is a result mainly of the absorbed solar radiation, and partly of radioactive decay within the planet, and the effective radiating temperatures of planets are shown in Fig. 1.2.

The rate at which solar energy reaches the outer limits of the Earth's atmosphere is known as the solar constant, and is estimated to be approximately 246 CHU/h-ft^2 ($5{\cdot}028 \times 10^6$ J/h.m^2). In space, this radiation intensity varies inversely with the square of the distance from the Sun. Energy is absorbed within the atmosphere, however, and at the Earth's surface the intensity does not exceed about 75 per cent of the solar constant.

At the Earth's distance from the Sun of 93 million miles (=one astronomical unit), the equilibrium temperature reached by a flat plate normal to the Sun's radiation would range from about 50° C upwards

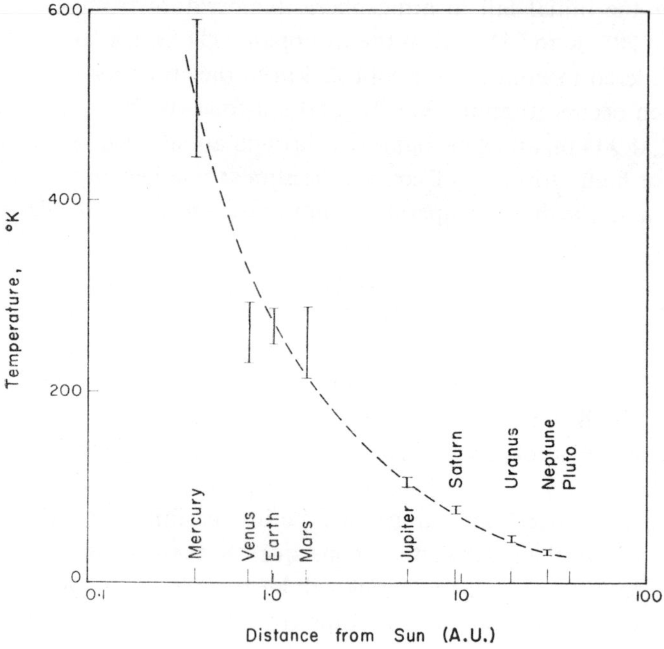

FIG. 1.2. Ranges of probable surface temperatures of planets (derived from refs. 14 and 18).

to an extent depending upon the variation of the absorptivity/emissivity ratio from unity, and on the emissivity of the back of the plate. The conclusion from this section, therefore, is that extreme conditions of temperature and of low pressure will be encountered in space, and that propulsion systems must be designed to maintain the reactant and propellant in their required states until consumption. As it happens, a large proportion of both materials will be utilized to escape from the Earth's atmosphere, hence volatile high-energy chemical reactants may be used for launch purposes but may involve serious storage difficulties if carried for return flight. They are less likely to pose major problems during re-entry since they will either have been consumed or, if in use for a powered descent, will not be subjected to such extremes of kinetic heating.

The most significant action-at-a-distance forces are those due to gravitational attraction between bodies in space. These forces vary inversely as the square of the distance between the centres of attraction; hence, for a planet of radius R_s, the local value of gravitational acceleration at a distance R_L from the centre of the planet is given by:

$$g_L = g_s R_s^2 / R_L^2 \tag{1.1}$$

where g_s is the gravitational acceleration at the surface of the planet. Since distances between planets are vast, the gravitational acceleration of one planet tends to be dominant at a given point in space.

1.3. Jet Thrust

The generation of thrust by the ejection of matter can be expressed mathematically by the application of the following Newton's laws of motion;

(i) Force = time rate of change of linear momentum (2nd law);
(ii) every action is associated with an equal and opposite reaction (3rd law).

From the former, it follows that the force involved in the displacement of matter may be expressed generally as;

$$F = \frac{d(mv)}{dt} = (\dot{mv}) = \dot{m}v + m\dot{v}.$$

In the particular case where velocity is constant and the time rate of change refers to mass, as with ejection of a stream of fluid, the expression becomes $F = \dot{m}v$. In the case where mass is constant and the velocity changes, as with acceleration of a rigid body, the expression becomes $F = m\dot{v}$. The application of the two laws to the case where a rigid body of constant mass M ejects a constant mass flow rate $\dot{m}$ of fluid at constant velocity v (relative to the body) gives the following:

Action in direction of ejection = Jet force = $\dot{m}v$,
which is numerically equal to;
Reaction in opposing direction = Thrust = $M\dot{V}$,

where $\dot{V}$ is the resultant acceleration of the rigid body if it is free to move.

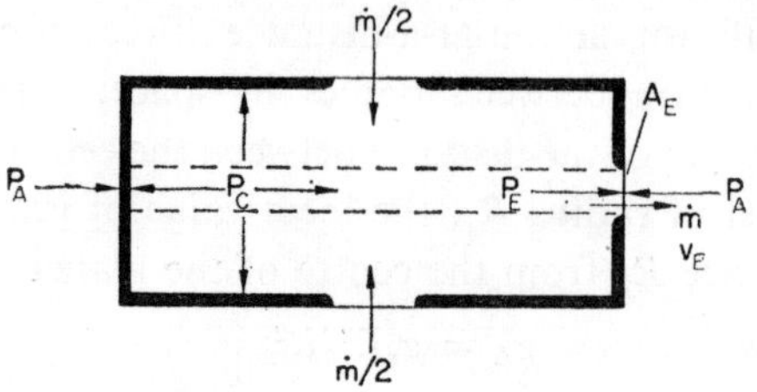

(a) Subsonic flow

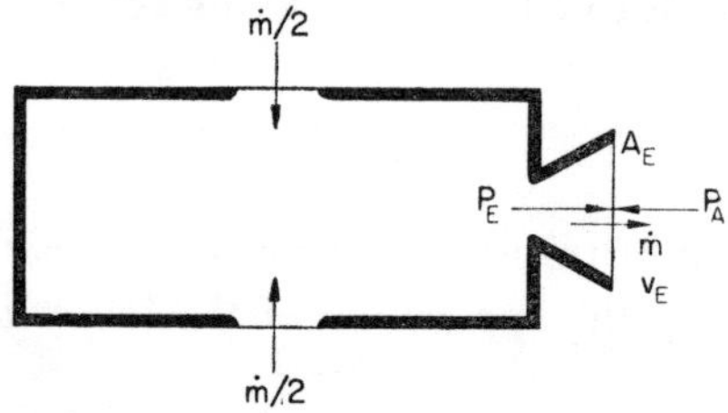

(b) Supersonic flow

FIG. 1.3. Momentum chamber.

Hence, numerically,

$$\text{Thrust} = F = M\dot{V} = \dot{m}v,$$

and vectorially,

$$M\dot{\mathbf{V}} = -\dot{m}\mathbf{v}, \tag{1.2}$$

to allow for the opposed directions.

This reasoning of jet thrust may be illustrated by means of the simple momentum chamber shown in Fig. 1.3a. Attention is drawn to conditions obtaining parallel to the horizontal chamber axis and, in particular, to those located within the cylinder of projected exit area, since all axial forces outside this cylinder are in equilibrium. The chamber contains compressible fluid at pressure P_c and, when the level of P_c is greater than that of the external pressure P_A, the fluid is constrained to flow axially through the exit plane A_E at a velocity v_E and mass flow rate $\dot{m}$. Replacement fluid enters the chamber perpendicularly to the axis at an equal mass flow rate $\dot{m}$ to maintain the pressure P_c in the chamber. Fluid within the chamber may be considered to be at rest, but fluid flowing through the exit plane possesses velocity and hence kinetic energy. From Bernoulli's law, therefore, the pressure energy of the exit flow is lower than that of the chamber fluid, hence static pressure P_E is less than P_c. In fact, in the subsonic efflux case in Fig. 1.3a, $P_E = P_A$, since any other value of P_E would result in either an expansion or contraction after the exit plane which would increase further the difference between P_E and P_A. Under these circumstances an internal axial force acts on the exit fluid as a momentum force numerically equal to $\dot{m}v_E$ and, in the opposite direction, as a momentum thrust $= (P_c - P_E)A_E = (P_c - P_A)A_E$ which is the difference between the internal and external forces acting on the leading surface of the projected cylinder.

The above conditions hold for subsonic exit flow where the internal pressure P_c is only moderately in excess of P_A. At a critical ratio of P_c/P_A, however, the exit velocity becomes sonic and the exit area chokes. For air, this critical pressure ratio is 1·895. Any further increase in P_c/P_A has no effect on the exit velocity v_E but increases the exit flow density and, in consequence, both the mass flow rate $\dot{m}$ and the thrust F. This condition in known as overcritical.

In Fig. 1.4 the mass flow rate, exit velocity and thrust have been plotted non-dimensionally in the form of ratios with corresponding values for the choked orifice, the choked values being represented by the superscript *. Curve A in the figure represents the constant, sonic, exit velocity through the choked orifice of Fig. 1.3a, and curve B shows the rising ratios of mass flow and thrust in the overcritical region. However, higher thrust is gained more effectively by raising, in addition, the v_E term in the thrust expression by means of supersonic diffusion through a divergent nozzle (Fig. 1.3b). Optimally, the new exit area A_E increases with P_c/P_A such that $P_E = P_A$ at all times. The mass flow varies as before, but the outlet velocity increases as shown by curve C. Being a product of $\dot{m}$ and v_E, thrust shows an added rise over the non-diffusion case, represented by the increment of curve D over curve B. These gains are substantial and, for this reason, supersonic diffusion is standard practice in rocket engine design.

The form of the P_c/P_A plot in Fig 1.4 could be applied directly to the case of variable P_c with constant P_A, analogous to a throttle-controlled rocket craft operating either at constant altitude or in space, provided the nozzle exit area varied with P_c to maintain P_E constant and equal to P_A. Equally important in the initial and final stages of a mission is the case of variable P_A with constant P_c, analogous to the craft changing altitude at fixed-throttle setting, consequently values of P_A/P_c are also indicated in Fig. 1.4 for optimum nozzle exit area.

In practice, the complication of variable-nozzle geometry is rarely justifiable, and a mean design must be employed over the range of

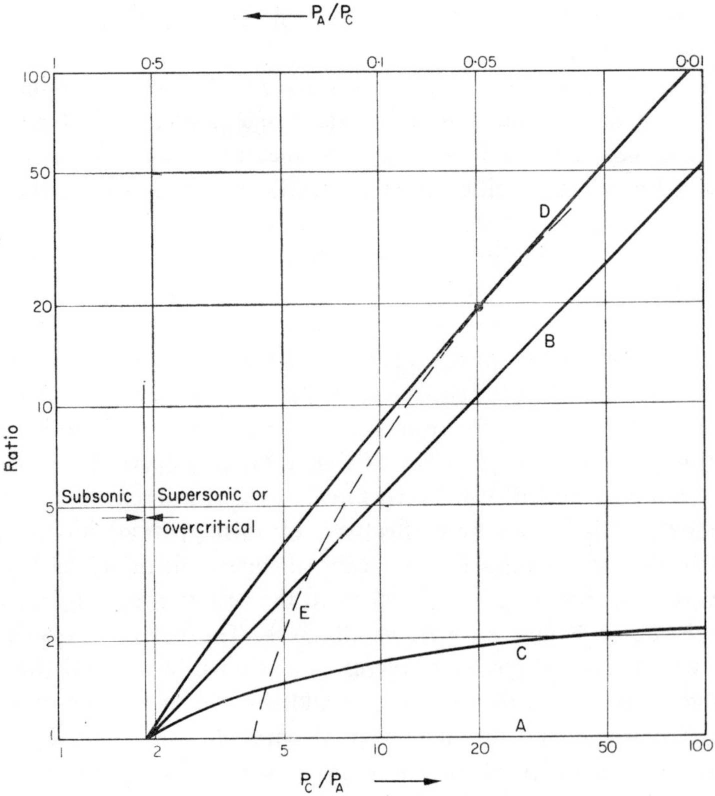

FIG. 1.4. Optimum nozzle performance curves.

operation. In theory, an off-design condition represents non-equality of P_E and P_A, with either an expansion or a shock wave downstream of the nozzle exit in order for the exit fluid to achieve the P_A pressure level virtually instantaneously. In such cases, a lower momentum thrust obtains and an additional external pressure force acts on the momentum chamber, giving a reduced total thrust, as follows:

Total off-design thrust $= F =$ momentum thrust + pressure thrust $= \dot{m}v_E + (P_E - P_A)A_E$. The total off-design thrust, therefore, is dependent upon ambient pressure P_A, and thus upon altitude. Operation in a vacuum, where $P_A = 0$, gives;

$$F = \dot{m}v_E + P_E A_E.$$

These off-design cases are referred to as underexpansion and overexpansion respectively. Underexpansion results from insufficient exit area such that P_E exceeds P_A, causing expansion to continue externally in a series of rarefaction discontinuities comprising a Prandtl–Meyer expansion wave (Fig. 1.5). Overexpansion resulting from excessive exit area tends to lower P_E below P_A, and pressure balance at the exit

Keyto Fig. 1.4.

Curve	Orifice with overcritical flow	Divergent nozzle with supersonic flow
A	V_E/V_t^*	—
B	$\dot{m}/\dot{m}^*$	$\dot{m}/\dot{m}^*$
	F/F^*	—
C	—	V_E/V^*, (optimum A_E/A_t)
D	—	F/F^* (optimum A_E/A_t)
E	—	F/F^* with A_E/A_t designed for $P_c/P_A = 20/1$

Superscript* represents conditions when orifice is choked.

area is achieved by compression shock waves. With moderate overexpansion, oblique shock waves form externally, but increasing levels of overexpansion bring shock-wave separation inside the nozzle, with progressive advancement towards the throat (Fig. 1.5). In each case, thrust is reduced from the design value. Included in Fig. 1.4 is a specific example of a nozzle designed to operate at a P_c/P_A ratio of 20/1, the broken line of curve E showing how the thrust ratio falls away from optimum with variation in P_c/P_A from the design value.

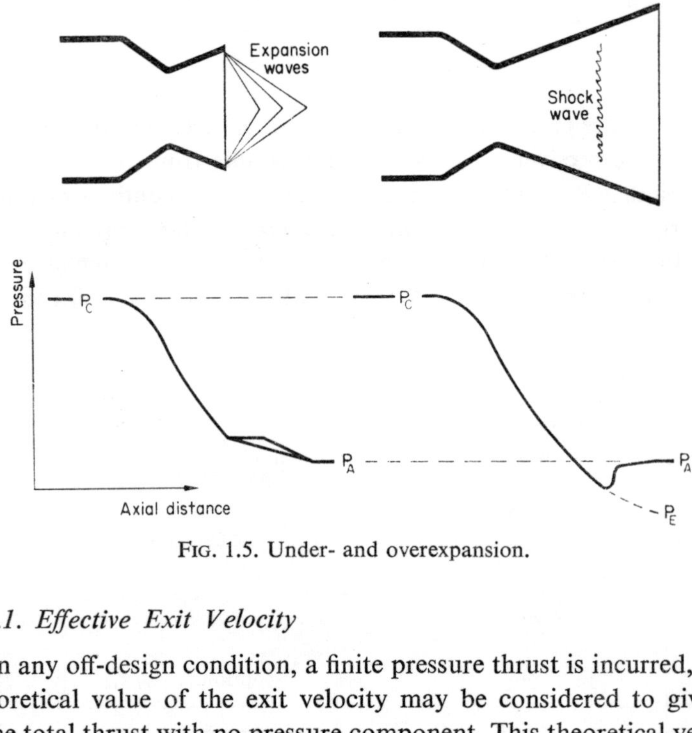

FIG. 1.5. Under- and overexpansion.

1.3.1. Effective Exit Velocity

In any off-design condition, a finite pressure thrust is incurred, but a theoretical value of the exit velocity may be considered to give the same total thrust with no pressure component. This theoretical velocity is known as the effective exit velocity, denoted by the symbol v_e,

i.e.

$$F = \dot{m}v_E + (P_E - P_A)\,A_E$$

$$F = \dot{m}v_e. \tag{1.3}$$

At all conditions of operation, therefore,

$$v_e = F/\dot{m} = \frac{\text{actual thrust}}{\text{actual mass flow rate of propellant}}$$

and, like F, is dependent upon the ambient pressure P_A.

1.3.2. Specific Impulse

Provided consistent units are used, the value of v_e is numerically equal to that of the specific impulse of the propellants, defined as the thrust produced per unit mass flow rate of propellants:

$$\text{Specific impulse} = I_{sp} = F/\dot{m} = v_e, \text{ in units of force}\times\text{time/mass,}$$

e.g. lbf.s/slug, or pdl-s/lbm (= lbf.s.ft/lbf.s^2 = ft/s).

It is frequently more convenient to deal with thrust in one system of units (lbf) and with propellant mass in related units in another system (lbm), and specific impulse is also commonly defined using the mixed lbf-lbm-ft-s system of units, as follows;

$$I_s = F/\dot{m}, \quad \text{lbf.s/lbm, by "mixed units" definition.}$$

In such cases, $I_s = I_{sp}/k = F/\dot{m}k = v_e/k$, where k is the numerical conversion factor between the two-unit systems, i.e. $k = 32{\cdot}174$ pdl/lbf, or lbm/slug.

A third method is to define specific impulse in terms of weight flow using consistent units, i.e.

$$I_w = F/\dot{w} = F/\dot{m}g, \quad \text{in units of second,}$$

where g is the gravitational constant = 32·174 ft/s^2. Thus the weight of the flowing propellant is related to international standard sea-level conditions on Earth, and not to the local value of gravitational acceleration in space. In the broad sense, this is a disadvantage, since propellant weight is being referred to one particular condition in the universe, whereas propellant mass is intrinsic and applicable anywhere (except at a velocity approaching that of light). On the other hand, expression in units of time only gives results for I_w common to all systems employing the second as the time unit. Furthermore, since $k = g$ numerically, values of I_s and I_w are also common in English unit systems (see Table 1.1).

TABLE 1.1. Summary of Propulsion Parameters

Parameter		Expression	English units	SI units
Thrust, F		$\dot{m}v_e$	lbf	N
Specific impulse	I_{sp}	$F/\dot{m} = v_e$	ft/s	m/s
	I_s	$[F/\dot{m}] = F/[k]\dot{m} = v_e/[k]$	lbf.s/lbm	—
	I_w	$F/\dot{w} = v_e/g$	s	s
Efflux power	P_e	$\frac{1}{2}\dot{m}v_e^2$	ft.lbf/s	W
Vehicle kinetic power		FV	ft.lbf/s	W

[ ] = mixed units.
$[k]$ = conversion factor between unit systems = 32·174 pdl/lbf.
g = gravitational constant = 32·174 ft/s² = 9·81 m/s².

Since a solid charge of reactant material is consumed at a mass rate dependent upon the total area exposed to reaction and not a rate of feed into the chamber, specific impulse can be derived from the total impulse which is defined as the integral of the thrust with respect to time, i.e.

$$\text{Total impulse} = I_T = \int_0^{t_T} F\,dt = F_M t_T,$$

where t_T = total reaction time, and F_M = mean value of thrust.

$$\text{Thus, mean } I_{sp} = F_M/\dot{m} = F_M t_T/m_T = I_T/m_T = \frac{\text{total impulse}}{\text{total reactant mass}}.$$

Hence, specific impulse is the impulse resulting from the consumption of unit mass of reactant.

Exit velocity, and hence specific impulse, may be derived in more fundamental terms from the principle of conservation of energy for an adiabatic flow process by applying the Bernoulli expression to the inlet (c) and exit (e) planes of the propelling nozzle, as follows;

$$\begin{aligned}\text{Total energy} &= \text{kinetic energy} + \text{enthalpy}\\ &= \tfrac{1}{2}mv_c^2 + mh_c, \quad \text{at inlet}\\ &= \tfrac{1}{2}mv_e^2 + mh_e, \quad \text{at exit, assuming no losses in the nozzle,}\end{aligned}$$

where h = enthalpy per unit mass, the two energy quantities being in dimensionally homogeneous units.

Since v_c is nominally zero, rearrangement of the expression is possible to give;

$$v_e = \sqrt{2(h_c - h_e)}.$$

The level of operating temperature in rocket reaction chambers is such that extensive dissociation, and possibly ionization, are likely, although the residence time within the nozzle may be sufficient for subsequent recombination such that chemical equilibrium shifts continuously with the changing proportions of components. In the theoretically ideal case, however, the chemical equilibrium established within the reaction chamber does not shift during flow through the nozzle. With the equilibrium frozen in this way, and a mean value of c_p taken over the range of operating temperature,

$$v_e = \sqrt{2c_pT_c(1 - T_e/T_c)}.$$

Furthermore, from the thermodynamic relationships,

$$T_e/T_c = (P_e/P_c)^{\frac{\gamma-1}{\gamma}}; \quad c_p - c_v = R; \quad c_p/c_v = \gamma;$$

and hence $\quad c_p = R\gamma/(\gamma-1) = G\gamma/\boldsymbol{m}(\gamma-1),$

$$v_e = \sqrt{\frac{2\gamma GT_c}{(\gamma-1)_m}\left[1 - (P_e/P_c)^{\frac{\gamma-1}{\gamma}}\right]} \tag{1.4}$$

where $\boldsymbol{m}$ is the mean molecular weight of the exit gases.

It has been shown that the effective exit velocity is the value of the actual exit velocity at the optimum condition where the pressures at nozzle exit and external field are equal. Both the effective exit velocity and the specific impulse are used widely to indicate performance of rockets and propellants but, in using these parameters, it is necessary to quote both the reaction chamber pressure and the ambient pressure that apply. It is helpful, therefore, to devise a performance parameter which is independent of ambient conditions.

1.3.3. Characteristic Exit Velocity

Since the overall thrust of a rocket engine is dependent upon the combined effects of the combustion chamber and the nozzle, it is logical to attempt to separate the contributions of these two components. The expression for specific impulse derived earlier, therefore, may be expanded as follows;

$$I_{sp} = F/\dot{m} = v_e = v'_e C_f, \quad \text{in consistent units,}$$

where v'_e is known as the characteristic exit velocity, and the non-dimensional term C_f is known as the thrust coefficient. A useful concept introduced here is the fictitious thrust that would result from the full chamber pressure acting at the nozzle throat. Although this has no physical significance, it is used to define the thrust coefficient, i.e.

$$\text{Actual thrust} = F = C_f\,(\text{Fictitious thrust}) = C_f(P_c A_t),$$

where A_t is the area of the nozzle throat. Hence, C_f represents the augmentation by the nozzle of this fictitious thrust, and normally ranges from 1 to about 2·2.

$$\text{From the above expression, } v'_e = F/\dot{m}C_f = F(P_c A_t)/\dot{m}F = P_c A_t/\dot{m},$$

and represents the specific impulse equivalent to the fictitious thrust at the throat. From the following thermodynamic relationships:

$$\dot{m} = \varrho_t A_t v_t = P_t A_t v_t / R T_t$$

$$v_t = \sqrt{\gamma R T_t} = \text{sonic velocity}$$

$$P_t = P_c\left[\frac{2}{\gamma+1}\right]^{\frac{\gamma}{\gamma-1}}, \quad \text{and} \quad T_t = T_c\left[\frac{2}{\gamma+1}\right], \quad \text{since Mach No. at}$$

throat = 1,

$$v_e' = \frac{P_c A_t}{\left(\frac{P_t A_t}{RT_t}\right)\sqrt{\gamma RT_t}} = \frac{P_c RT_t}{P_c\left[\frac{2}{\gamma+1}\right]^{\frac{\gamma}{\gamma-1}}\sqrt{\gamma RT_t}} = \sqrt{\frac{RT_t}{\gamma}}\left[\frac{\gamma+1}{2}\right]^{\frac{\gamma}{\gamma-1}}$$

$$= \sqrt{\frac{RT_c}{\gamma}}\left[\frac{\gamma+1}{2}\right]^{\frac{\gamma+1}{2(\gamma-1)}}$$

Being a function of T_c and γ only, v_e' meets the requirements of a parameter which depends solely upon the properties of the products of the reaction chamber, and is independent of ambient conditions. Hence, v_e' represents a combined characteristic of the propellant and chamber performance, and is influenced by the energy content and consumption rate of the reactants, together with the design of the chamber. It is used primarily with chemical rocket engines.

Similar treatment of C_f shows it to be a function of γ, ratio P_E/P_c, ambient pressure P_A, and nozzle area ratio A_E/A_t. Since γ is involved, C_f does have some propellant dependence, but it serves mainly to collate the influence of the nozzle and of ambient pressure upon the total thrust.

1.3.4. The Rocket Equation

The basic vectorial thrust relationship (equation (1.2)) may be solved for the velocity of the rocket vehicle, as follows:

$$F = M\frac{dV}{dt} = -v\frac{dm}{dt},$$

therefore, $dV = -v\,dm/M,$

$$= -v\,dM/M,$$

since, for a rocket, the propellant is entirely vehicle-borne. Integration between initial and final conditions gives;

$$\int_{V_i}^{V_f} dV = -v\int_{M_i}^{M_f} dM/M,$$

i.e. the velocity increment $= V_f - V_i = \Delta V = v \ln M_i/M_f$.

This basic rocket equation may be written in more general terms to cover off-design conditions, as follows;

$$\Delta V = v_e \ln M_i/M_f. \tag{1.5}$$

Hence, the vehicle velocity increment resulting from the consumption of propellants of mass $(M_f - M_i)$ is proportional to the effective propellant exit velocity v_e, and the ratio of the initial to final vehicle mass, M_i/M_f. In the particular case where the initial velocity is zero, as in a static ground launch, and where the reactant is utilized continuously to completion, the expression gives the burnout velocity of the vehicle as:

$$V_b = v_e \ln M_i/M_b,$$

where M_b is the burnout mass of the vehicle.

For powered flight against a gravitational force, the thrust equation (1.2) is modified by the vehicle weight, Mg, as follows:

$$F = M\frac{dV}{dt} = -v\frac{dM}{dt} - Mg_M,$$

where g_M is the mean value of the gravitational acceleration.

Hence,
$$dV = -v\frac{dM}{dt} - g_M\,dt,$$

and,
$$V_b = v_e \ln M_i/M_b - g_M t_T,$$

where t_T = total reaction time. Following a static ground launch, the reduction in burnout velocity due to gravity may reach a level of about 30 per cent. A further reduction of about 10 per cent may result from the aerodynamic drag through the Earth's atmosphere.

By anticipating the efflux velocities given in later chapters for various propulsion systems, curves of velocity increment versus vehicle mass ratio can be drawn as in Fig. 1.6 using the rocket equation. Superimposed on the diagram are the velocities (not velocity increments) required at the Earth's surface both for a circular orbit and for escape. Typical figures for the payload and structure mass of a chemical-rocket vehicle are 5 per cent and 9 per cent respectively of the initial, fully-loaded, mass of the vehicle, giving a M_i/M_f mass ratio of 7·15. The rocket equation shows that such a vehicle launched from rest in the absence of gravitational and aerodynamic losses would achieve a burnout velocity of 26,000 ft/s, which is sufficient for the surface circular orbit (see Fig. 1.10).

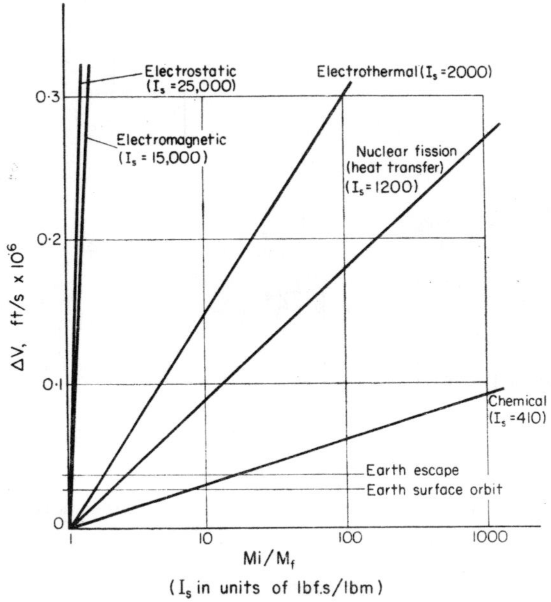

FIG. 1.6. Rocket vehicle velocity increments (I_s in units of lbf.s/lbm).

By constructing the vehicle in a number of self-contained stages, the structure of each burntout stage can be jettisoned in turn such that the mass of the remaining structure is more closely scaled to the mass of the remaining propellant, i.e. the rising level of structure/propellant mass ratio is displaced back towards its initial value. The need for, and effectiveness of, vehicle staging can be illustrated as in Fig. 1.7. If twice the above velocity is required, i.e. 52,000 ft/s, it may be achieved with a two-stage vehicle either at the expense of the payload, i.e. 0·25

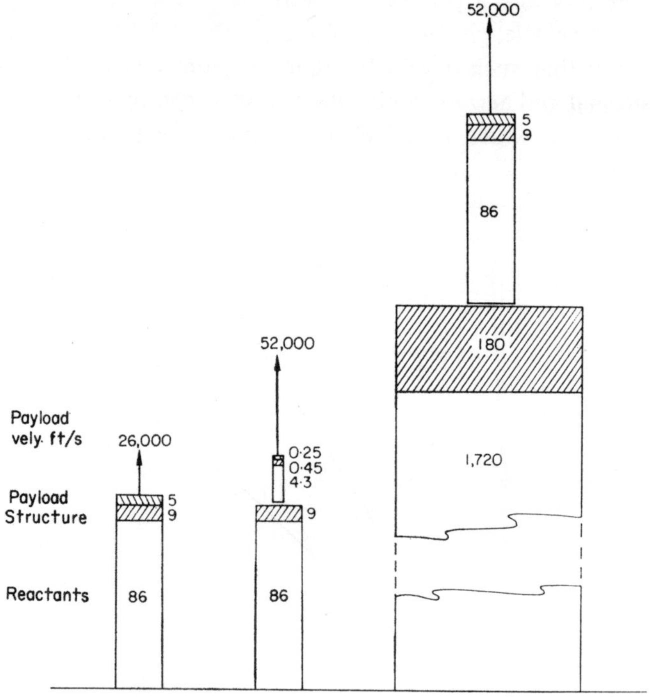

Number of stages	One	Two	Two (large scale)
Payload vely. ft/s	26,000	52,000	52,000
Payload mass	5	0·25	5
Initial mass	100	100	2000

FIG. 1.7. Two-stage performance.

instead of 5 mass units, or at the expense of initial mass, i.e. 2000 instead of 100 mass units. Additional stages, up to a maximum of about ten, raise the final payload velocity but at further expense of payload and/or initial mass, together with added complication and cost to achieve the required reliability.

1.3.5. Alternative Systems of Jet Acceleration

The conventional method of achieving high thrust from a rocket chamber is to drive the propellant up to sonic velocity at the nozzle throat in order to promote supersonic acceleration in the nozzle divergence. To date, the driving pressure has been evolved by chemical combustion, and the limiting parameter is the ratio of combustion temperature to product molecular weight, i.e. $T_c/\boldsymbol{m}$. Comparable driving pressures are likely with nuclear heating, and marked improvement in performance is possible due to the lower values of $\boldsymbol{m}$. Alternatively, the energy may be added to the propellant in the chamber by an electrical arc discharge, or by resistance heating.

Jet thrust may also be achieved by the use of an electrically-conducting propellant accelerated by means of electrostatic or electromagnetic forces, or alternatively by the emission of photons. All these systems are examined in later chapters.

1.3.6. Efficiencies of Energy-thrust Conversion

An expression for efficiency can be derived for each step in the conversion process of stored energy to thrust.

In the chemical rocket, combustion efficiency can be defined as the ratio of the actual change in enthalpy per unit mass of products to the theoretical change in enthalpy that would result from the ideal consumption of the reactants. Thus, combustion efficiency, η_H, is given by;

$$\eta_H = \frac{h_c - h_i}{Q_r} \quad \text{for the chemically-powered rocket,}$$

where h_c = enthalpy per unit mass of propellant after reaction, h_i = enthalpy per unit mass of reactant before reaction, and Q_r = net reaction energy per unit mass of reactants based on initial condition. Since the reactants are assumed to have no axial flow in the chamber during reaction, these h values are total or stagnation enthalpies. Experimental measurements are rarely made, but combustion inefficiencies through non-ideal combustion normally reach about 5 per cent.

When propellant mass flow rate is included in the combustion efficiency expression:

$$\eta_H = \frac{\dot{m}(h_c - h_i)}{\dot{m}Q_r} = \frac{\text{power absorbed by propellant}}{\text{power equivalent to reactants consumed}} = \frac{P_p}{P_r}.$$

When the propellant is heated independently, as from a nuclear reactor by means of a heat exchanger or an electric arc, this power ratio is known as the propellant heater efficiency, i.e.

$$\eta_H = \frac{P_p}{P_r} = \frac{P_p}{\dot{n}E_f},$$

for the nuclear-powered rocket, where $\dot{n}$ = rate of fission occurrence, and E_f = energy per fission. Here, the inefficiency is incurred by heat-transfer losses through the outer walls of the reaction chamber.

In the frozen-flow case, additional losses are experienced through the energies of dissociation and/or ionization retained in the efflux. These are dependent upon the properties of the propellant at the given energy level, and may be represented by the following expression for frozen-flow efficiency:

$$\eta_F = \frac{\text{kinetic power of efflux}}{\text{power absorbed by propellant}} = \frac{P_e}{P_p}.$$

Hence, for the reactant-engine system as a whole,

$$\text{Overall internal efficiency} = \eta_I = \frac{P_e}{P_r} = \eta_H \eta_F.$$

A propulsive efficiency, η_P, may also be derived for the engine-vehicle combination as the ratio of the kinetic power expended by the efflux in maintaining the vehicle velocity, to the total kinetic power of the efflux:

$$\eta_P = \frac{\text{kinetic power of vehicle}}{\text{kinetic power of vehicle} + \text{residual power of efflux}}.$$

When the velocity of a moving vehicle is achieved by means of a force applied to overcome resistance to motion, the work continuously being done is given by;

$$W = \text{resistance force} \times \text{distance moved} = R'x.$$

For a rocket vehicle travelling at constant velocity V under the action of a constant thrust F equal to resistance R',

$$\text{kinetic power of vehicle} = R'\frac{dx}{dt} = R'V = FV.$$

Since work done = kinetic energy $= \frac{1}{2}mv^2$, and since power is the rate of doing work, the kinetic power of the efflux, P_e, is $\frac{1}{2}\dot{m}v_e^2$, where $\dot{m}$ is the mass rate of propellant being energized, and v_e is the exit velocity. After ejection the efflux retains an absolute velocity equal to $(v_e - V)$, and thus a residual power equal to $\frac{1}{2}\dot{m}(v_e - V)^2$ which goes to waste.

Hence,
$$\eta_P = \frac{FV}{FV + \frac{1}{2}\dot{m}(v_e - V)^2},$$

and, since
$$F = \dot{m}v_e,$$

$$\eta_P = \frac{2V/v_e}{1 + (V/v_e)^2}. \tag{1.6}$$

The plot of propulsive efficiency vs. vehicle/efflux velocity ratio in Fig. 1.8 shows that 100 per cent efficiency obtains when vehicle and efflux velocities are equal, since the residual power of the efflux disappears, and also indicates that the velocity of a rocket vehicle may exceed that of its propellant efflux.

The performance chart for rocket vehicles with 100 per cent propulsive efficiency (Fig. 1.9) shows the relationship between the various parameters, together with the regions of high mass flow, efflux power and specific impulse. The ratio of thrust to efflux power evaluates as follows:

$$\frac{F}{P_e} = \frac{\dot{m}v_e}{\frac{1}{2}\dot{m}v_e^2} = \frac{2}{v_e} = \frac{2}{I_{sp}}.$$

Thus, an economy of propellant material results from an increase in specific impulse, but at the expense of the power required from the energy source to maintain a given thrust.

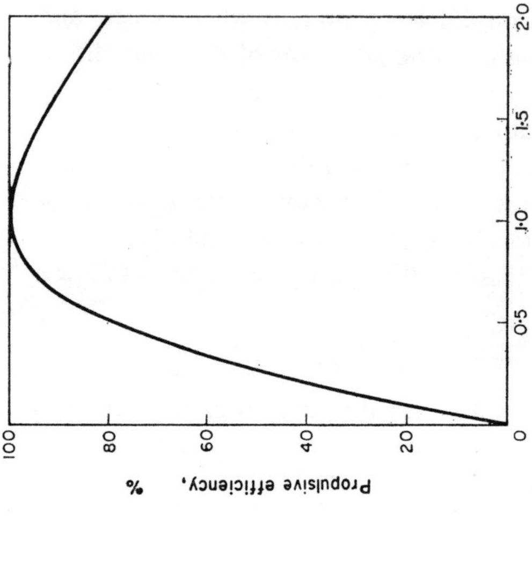

FIG. 1.8. Propulsive efficiency variation with vehicle/efflux velocity ratio.

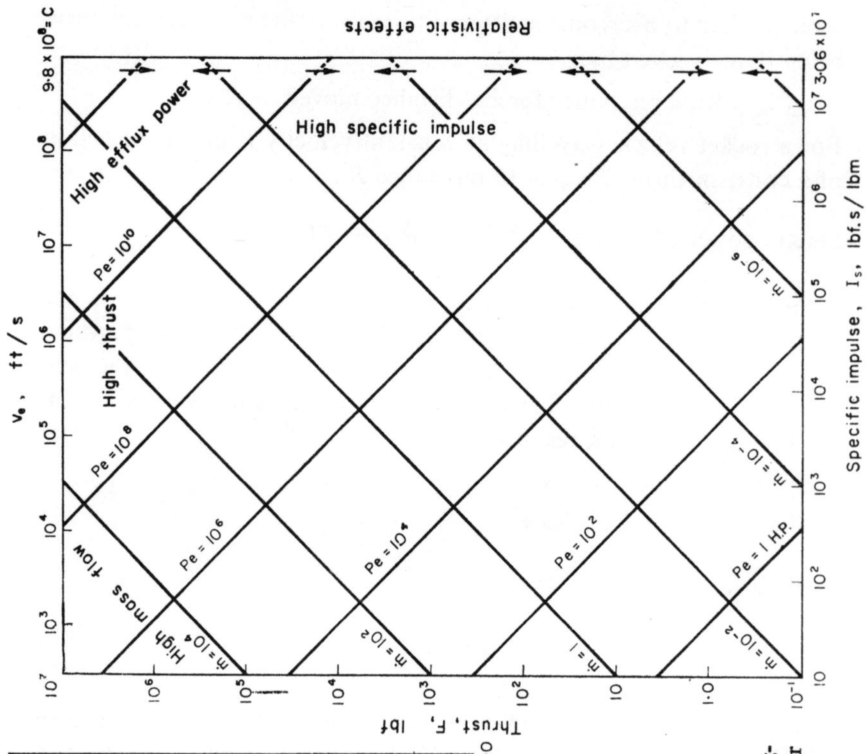

FIG. 1.9. Performance chart for rocket engines. (Propulsive efficiency assumed 100 per cent.) (Form of plot derived from ref. 22.)

1.4. Propulsion Requirements for Spaceflight

The prime requirement for flight within interplanetary space is that the vehicle shall have sufficient energy to negotiate the gravitational field obtaining between the launch and target planets. Initially, this means that the vehicle must overcome atmospheric drag and approach the escape condition from its launch planet, the closeness of this approach being dependent upon the distance of the target planet.

A vehicle travelling in space thus possesses potential energy in respect of its distance from the principal gravitating body, together with kinetic energy related to its velocity.

The potential energy of the vehicle is equal to the quantity of work expended in climbing from the surface of the launch planet to the point in space. That is,

Potential energy $= PE = M \int_{R_s}^{R_L} g_L \, dR_L = Mg_s R_s^2 \int_{R_s}^{R_L} dR_L$, from equation (1.1)

$$= Mg_s R_s^2 \Big[-1/R_L\Big]_{R_s}^{R_L} = Mg_s R_s^2 \left(\frac{1}{R_s} - \frac{1}{R_L}\right)$$

$$= Mg_s R_s \left(1 - \frac{R_s}{R_L}\right),$$

where M = mass of vehicle.

The kinetic energy of the vehicle is a function of the linear velocity, and is independent of the direction. That is,

$$\text{Kinetic energy} = KE = \tfrac{1}{2} MV^2.$$

Ignoring all other forms of stored and radiated energy, therefore, the total energy of the vehicle may be expressed as;

$$\text{Total energy} = E = PE + KE = Mg_s R_s \left(1 - \frac{R_s}{R_L}\right) + \tfrac{1}{2} MV^2.$$

For unit mass of vehicle,

$$E/M = g_s R_s \left(1 - \frac{R_s}{R_L}\right) + \tfrac{1}{2} V^2.$$

Since the gravitational acceleration of one planet is likely to be dominant, the path of a space vehicle in free fall takes the form of a conic whose focus lies at the centre of the principal gravitating body. The particular form of conic followed depends mainly upon the linear velocity of the vehicle, and partly upon its altitude. Under conditions where the centrifugal force exerted on the vehicle just balances the gravitational force, the path takes a circular form around the body centre, i.e.

$$\frac{MV_o^2}{R_L} = Mg_L = Mg_s \frac{R_s^2}{R_L^2}$$

where V_o = circular orbit velocity.

Hence, $V_o = R_s \sqrt{\frac{g_s}{R_L}}$, and in terms of altitude H,

$$V_o = R_s \sqrt{\frac{g_s}{R_s + H}}.$$

For Earth, the variation of V_o with H is shown in Fig. 1.10.

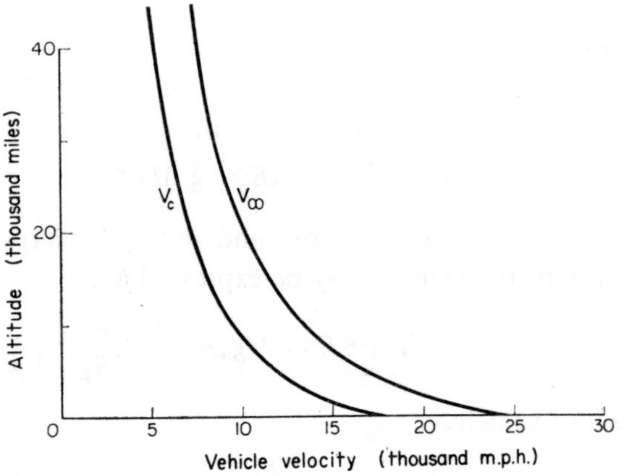

FIG. 1.10. Variation of circular-orbit and escape velocities with altitude (Earth).

At the circular orbit condition,

$$\frac{KE_o}{M} = \tfrac{1}{2}V_o^2 = \frac{g_s R_s^2}{2R_L}.$$

Hence, $$\frac{E_o}{M} = g_s R_s\left(1-\frac{R_s}{R_L}\right)+\frac{g_s R_s^2}{2R_L} = g_s R_s\left(1-\frac{R_s}{2R_L}\right). \qquad (1.7)$$

With increasing velocity at a given altitude, the path takes the form of an ellipse of increasing major axis until, at the escape velocity, V_∞, the path opens out into a parabola and the vehicle moves continuously away from the body. The escape velocity of the vehicle located at a given radius from a planet can be calculated from the kinetic energy required to be expended as work in transferring the vehicle from that radius to infinity against the gravitational force of the planet, i.e.

$$KE_\infty = \tfrac{1}{2}MV_\infty^2 = M\int_{R_L}^{\infty} g_L\,dR_L = Mg_s R_s^2\int_{R_L}^{\infty}\frac{dR_L}{R_L^2} = \frac{Mg_s R_s^2}{R_L}$$

Thus, $$V_\infty = R_s\sqrt{\frac{2g_s}{R_L}} = V_o\sqrt{2}$$

and, in terms of altitude H, $V_\infty = R_s\sqrt{\dfrac{2g_s}{R_s+H}}$

For Earth, the variation of V_∞ with H is included in Fig. 1.10.

At the escape condition,

$$\frac{KE_\infty}{M} = \tfrac{1}{2}V_\infty^2 = \frac{g_s R_s^2}{R_L}$$

Hence, $$\frac{E_\infty}{M} = g_s R_s\left(1-\frac{R_s}{R_L}\right)+g_s R_s^2 = g_s R_s. \qquad (1.8)$$

For Earth, $g_s = 9{\cdot}8\ \text{m/s}^2 = 32{\cdot}174\ \text{ft/s}^2$,

and $R_s = 6{\cdot}38\times10^6\ \text{m} = 20{\cdot}91\times10^6\ \text{ft}$.

Hence, $$\frac{E_\infty}{M} = 6{\cdot}25\times10^7\ \text{m}^2/\text{s}^2,\ \text{or } J/\text{kg},$$

$$= 6{\cdot}73\times10^8\ \text{ft}^2/\text{s}^2,\ \text{or ft-lbf/slug}.$$

Since $E_\infty = (Mg_s)R_s$, the total energy for escape can be visualized as the work required for the vehicle to climb to an altitude equal to the radius of the planet, assuming a constant surface value of gravitational force throughout the climb. Values of escape energy for the planets are listed in Table 1.2.

TABLE 1.2. Planetary Escape Energies

Planet	Distance from Sun (a.u.*)	g_s Rel. Earth	R_s (Mile$\times 10^3$)	E_∞/M (J/kg)
Sun	0	28	432·5	$1{\cdot}91\times10^{11}$
Mercury	0·39	0·26	1·05	$8{\cdot}74\times10^{6}$
Venus	0·73	0·90	3·80	$5{\cdot}24\times10^{7}$
Earth	1	1·00	3·96	$6{\cdot}25\times10^{7}$
Moon	1	0·16	1·08	$2{\cdot}81\times10^{6}$
Mars	1·54	0·38	2·1	$1{\cdot}25\times10^{7}$
Jupiter	5·2	2·65	42·5	$1{\cdot}77\times10^{9}$
Saturn	9·6	1·14	35·0	$6{\cdot}28\times10^{8}$
Uranus	19·4	0·96	15·6	$2{\cdot}44\times10^{8}$
Neptune	30	1·0	16·5	$3{\cdot}08\times10^{8}$
Pluto	39·3	?	?	$5{\cdot}0\times10^{7}$

* 1 a.u. = 1 astronomical unit = 93×10^6 mile.

Energy data reproduced by permission from Corliss, *Propulsion Systems for Space Flight*, McGraw Hill, 1960 (ref. 10).

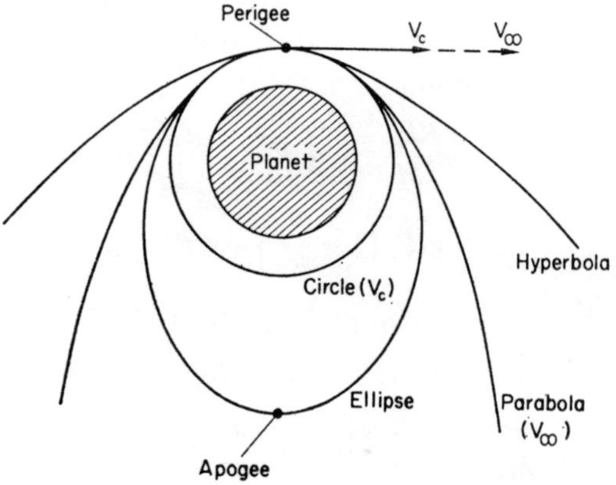

FIG. 1.11. Vehicle trajectories relative to planet.

At still higher velocities, the vehicle follows a hyperbolic path in its escape from the gravitational influence of the body (Fig. 1.11).

At altitude represented by radius R, values of PE/M, E_o/M, and E_∞/M are plotted for the Earth–Moon system in Fig. 1.12. Since the full line represents potential energy, this type of diagram is generally referred to as a potential-energy well for the planet concerned. In this

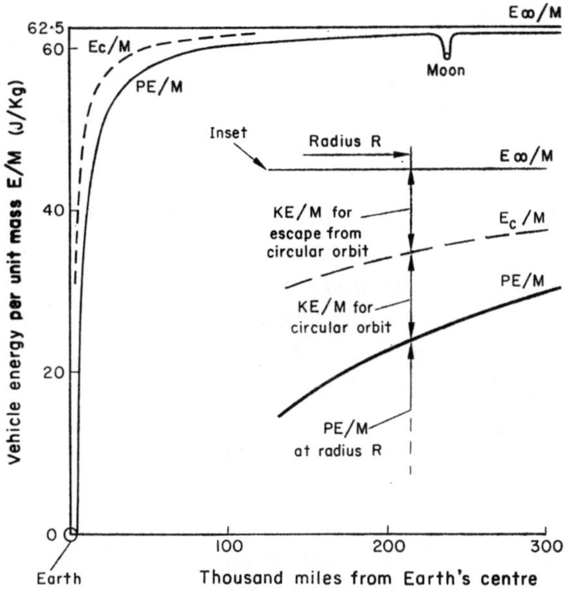

FIG. 1.12. Vehicle energy diagram for Earth–Moon system (ref. 13).

case, the potential-energy well for the Moon has been superimposed on that of the Earth. The broken line represents the total energy of the vehicle when in circular orbit, the kinetic energy involved being half that required for escape. The figure shows clearly the need for a large proportion of energy during the initial stages of the mission.

The more general presentation of Fig. 1.12 is noteworthy in which the zero is located at the escape level at the top of the well, with negative quantities of energy down to (in this case) $-62{\cdot}5$ J/kg at the bottom of the well. The reasoning is that, to an external observer, the energy stock of a system of two bodies with infinite separation normally appears to be zero, hence any approach of the bodies signifies a release of energy and a corresponding negative overall stock. The special plotting of Fig. 1.12 is permissible here, however, since the observer is, in fact, on the surface of the main body Earth, and takes the energy datum when the two bodies (Earth and vehicle) are in contact.

The larger scale inset of Fig. 1.12 permits a clearer understanding of the equilibria of satellites, and the apparent paradoxical effects when manoeuvring between orbits. The progressive operation of a retro rocket, for example, to a satellite in circular orbit would reduce its total stock of energy, bring it to a circular orbit of smaller radius where *PE* is less, and force it to travel at a higher velocity since the orbital *KE* is greater. In fact, a quantity of energy equal to the change in total energy is transferred from *PE* to become additional *KE*, thus a *braking* action results in an *acceleration*.

By terrestrial standards, the levels of velocity for orbit and escape are very high, and propulsion systems of very high performance are a natural prerequisite for exploration beyond the Earth's gravitational field. As events have shown, these levels of performance are already practicable achievements, but the successful development of manned spaceflight to distant planets, with adequate payload and sufficient resources for a safe return, calls for economy of propellants and a careful matching of propulsion system to mission requirements, i.e. to system optimization. Each manoeuvre comprising the complete mission will present its own requirements, and therefore lead to specific optimization. For example, a thrust/weight ratio in excess of unity is essential for launch purposes, whereas a comparatively low thrust would suffice for a continuous low acceleration in flight. A clearer indication of these requirements emerges from the consideration that thrust is composed of two factors, i.e. the mass flow rate and the velocity of the propellant efflux. Hence, optimization towards high efflux velocity (equivalent to specific impulse) leads to propellant economy. On the other hand, propellant mass and economy are less critical at launch, where the high thrust is essential. Thus, an appreciable propellant consumption can be tolerated throughout the launch, with a correspondingly lower specific impulse.

In summary, specific impulse is a measure of propellant enonomy, thrust is a measure of engine effectiveness, and thrust/weight ratio gives an indication of the price to be paid for this effectiveness in terms of engine weight.

For reference purposes in interplanetary travel, a minimum energy requirement is employed, and this can be shown to obtain for a "Hohmann" transfer ellipse, which is tangential to the solar orbits of both

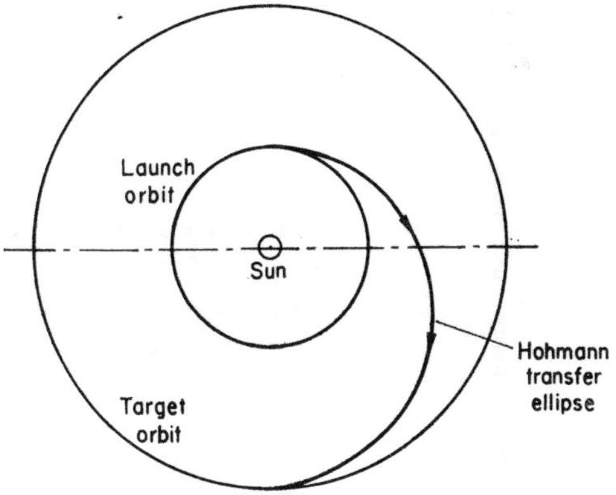

FIG. 1.13. Hohmann ellipse for minimum-energy transfer.

the launch and target planets, as in Fig. 1.13. The minimum total mission energy is then obtained, approximately, by adding the Hohmann transfer energy to the escape energies of the launch and target planets. All energy-changes must be added together, since the propulsion system supplies energy for thrust in both the forward and reverse directions.

Approximate values of total mission energy per unit mass of vehicle, calculated as above, are shown in Fig. 1.14 for one-way transit Earth to Venus, and Earth to Jupiter. These are minimum values, so that the requirements for shorter mission durations give corresponding increases in energy requirement levels. Return journeys would present corresponding requirements in addition, but on a smaller scale since the propellant mass to be carried, and probably the vehicle mass, would be less. The prime requirements for the various stages of an interplanetary mission are shown in Table 1.3, in terms of thrust (F), vehicle weight (W) and specific impulse (I_s).

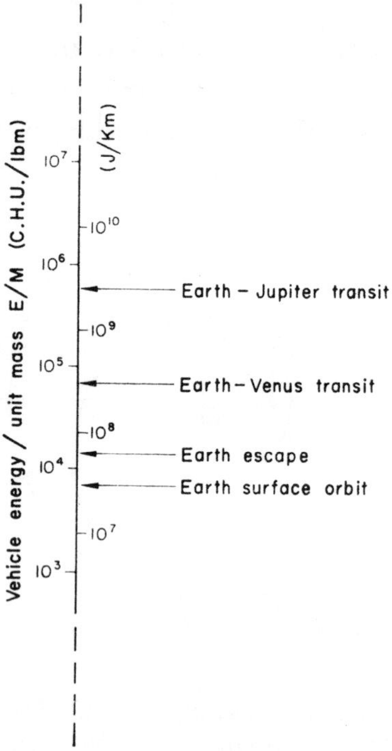

FIG. 1.14. Approximate minimum requirements for vehicle mission energy.

1.5. Energy Conversion for Thrust

Since energy is a prerequisite for motion, and since very high rates of movement are envisaged for spaceflight, highly concentrated sources of energy are essential. The rocket engineer, therefore, is interested in any and every reaction that promises plentiful energy release, and is prepared to seek far beyond the realms of conventional fuel combustion. In an examination of possible energy sources, the first subdivision

TABLE 1.3. Interplanetary Mission Requirements*(13)

Manoeuvre		Prime Requirements	
Launch to Earth orbit	1. Vertical launch 2. Sustained ascent 3. Entry to Earth orbit	$F/W > 1$, thus F_{max} $F \simeq 20\% \ F_{max}$	Min. I_s $\simeq 200$
Interplanetary transfer	4. Entry to transfer orbit 5. Entry to planetary orbit (retro)	$F \simeq 10\% \ F_{max}$	Min. I_s $\simeq 400$
Planetary orbit to touch-down	6. Entry to landing approach (retro) 7. Touchdown (retro)	$F \simeq 5\% \ F_{max}$, dependent on mass and escape vely. of target planet	
Orientation control		F approx. $1\% \ F_{max}$	

* A typical I_s value for practicable mission duration is 6000 lbf.s/lbm but peaks of 20,000 lbf.s/lm or more may be involved, depending upon the basis of the propulsion programme for the mission.

lies between those which are localized and can be stored aboard the vehicle, and those which exist broadly dispersed throughout the field of space in which the vehicle is to operate. These divisions are described as transportable and environmental respectively. The major energy sources and energy-releasing reactions of promise are dealt with generally in the next Chapter, and the various propulsive applications are covered separately in the following Chapters. Of the latter, transportable systems are considered first, since these are of interest currently and in the immediate future, whereas environmental systems of the more remote future are examined in a later Chapter.

In spacecraft propulsion, the common denominator in all transportable systems and most environmental systems is the expulsion of propellant material, hence the general objective is conversion of available energy to kinetic energy. It is rarely possible to effect this conversion in a single step and, in practice, rocket propulsion involves consecutive conversions of energy through a number of intermediate forms. Furthermore, several routes are possible from the available to the kinetic form. Since an energy loss accompanies each conversion process, highest overall conversion efficiency tends to follow the smallest number of conversion stages. However, operational limitations imposed by material properties can make some intermediate conversions attractive, as exemplified by the systems of electrical propulsion discussed in Chapter 5. All major propulsive systems included in this book are summarized in block diagram form in the final Chapter. The low-grade energy losses resulting from conversion inefficiencies collect within the vehicle and can be disposed of only by radiation to the heat sink of space, and this calls for careful and extensive design of radiation surfaces.

Symbols

A_E = Cross-sectional area of exit plane
A_t = Cross-sectional area of throat of supersonic nozzle
C_f = Thrust coefficient
c_p = Specific heat at constant pressure
c_v = Specific heat at constant volume
E = Total energy of vehicle $= PE + KE$
E_f = Energy per fission
F = Thrust
G = Universal gas constant
g = International gravitational constant
g_L = Local value of gravitational acceleration
g_S = Gravitational acceleration at surface of planet
H = Altitude
h = Enthalpy per unit mass of propellant
I_s = Specific impulse in units of lbf-s/lbm
I_{sp} = Specific impulse in consistent units
I_T = Total impulse
I_W = Specific impulse on weight flow basis
KE = Kinetic energy of vehicle

k = Numerical conversion factor between unit systems
M = Vehicle mass
m = Propellant mass
$\mathbf{m}$ = Mean molecular weight of propellant
m_T = Total mass of solid reactant
$\dot{n}$ = Rate of fission occurrence
P_A = External ambient pressure
P_E = Static pressure of propellant at nozzle exit plane
P_e = Kinetic power of efflux
P_p = Power absorbed by propellant
P_r = Power equivalent to reactants consumed
PE = Potential energy of vehicle
Q_r = Net reaction energy of reactants
R = Gas constant
R_L = Distance from centre of planet
R_S = Radius of planet
T_e = Static temperature of propellant at nozzle exit plane
t = Reaction time
V = Vehicle velocity
v = Propellant velocity
v_E = Velocity of propellant through nozzle exit plane
v_e = Effective velocity of propellant through nozzle exit plane
v_e' = Characteristic velocity of propellant through nozzle exit plane

γ = c_p/c_v
η_H = Efficiency of combustion, or of propellant heater
η_F = Frozen-flow efficiency
η_P = Propulsive efficiency
η_I = Overall internal efficiency
ϱ = Density of propellant

Superscripts

$*$ = Conditions when orifice choked

Subscripts

b = Burnout conditions
c = Reaction chamber conditions
f = Final conditions
i = Initial conditions
M = Mean value
o = Circular orbit conditions
T = Total
t = Nozzle throat conditions
∞ = Planetary escape conditions

CHAPTER 2

ENERGY RELEASE

AN UNDERSTANDING of the processes of energy release from material requires an appreciation of the structure of the material itself, since the energy results from changes in the structure. Two levels of sub-microscopic dimension are of interest here; changes in the atom as a whole and in its relationship with other atoms give rise to the emission of radiant energy and chemical energy respectively, whereas changes in the atomic nucleus lead to copious energy release through the annihilation of mass. These fundamental structural considerations are considered briefly in this chapter, together with a broad illustration of the associated sources of energy.

2.1. Atomic Structure

2.1.1. Electron Orbits

In general terms, an atom comprises a nucleus of protons and neutrons in which most of the mass is concentrated, and around which revolve electrons grouped in shells. For an atom in the normal, electrically-balanced, state, the numbers of protons and electrons are equal. The electrons revolve only in certain selected orbits called stationary states, each of which corresponds to a certain energy level. Electrons closest to the nucleus possess least energy.

The shells of orbits are assigned a principal quantum number from 1 to 7, and/or lettered *K* to *Q*, consecutively outwards, and the main orbit within them is lettered *s*. The 2 to 6, or *L* to *P*, shells also comprise groups of orbits which are lettered *p*, *d* and *f*, and divided further into

3, 5 and 7 sub-orbits respectively. Furthermore, a maximum of two electrons can occupy each orbit or sub-orbit provided they spin in opposite directions. The overall picture, therefore, appears as in Table 2.1.

The groups of orbits and sub-orbits shown above are assigned a subsidiary (or azimuthal) quantum number, denoted by the symbol l, which ranges from 0 for the *s*-orbit to 3 for the *f*-orbit. Since the *p*-orbits

TABLE 2.1. Arrangement of Electron Orbits

Shell:		*K*	*L*	*M*	*N*	*O*	*P*	*Q*
Principal quantum no. (= n)		1	2	3	4	5	6	7
Orbits	1	Ks	Ls	Ms	Ns	Os	Ps	Qs
Sub-orbits	3		Lp	Mp	Np	Op	Pp	
	5			Md	Nd	Od	Pd	
	7				Nf	Of		
Max. number of electrons in shell (= $2n^2$)		2	8	18	32	*	*	*

* Complicated by transitional nature of elements.

take up different positions under the action of a strong magnetic field, a magnetic quantum number, denoted by symbol m_l, is also necessary to define the orbit more precisely. The requirement for opposite spin leads to a spin quantum number, denoted by the symbol m_s, with values equal to either plus or minus $\frac{1}{2}$, hence an electron is completely defined by specifying the four quantum numbers appropriate to it.

Rutherford's model of an electron comprises a small particle traversing a closed orbit, but the modern concept is a cloud of negative charge somewhat similar to the result of a prolonged photographic exposure of a rapidly orbiting particle. The density of the photographic image at any point would bear a direct relationship with the probability of finding the particle at that point at any given instant. On this basis, the charge cloud in the hydrogen atom appears as a sphere with maximum probability density at a small radius from the nucleus due to the attraction of the proton. The lithium atom exhibits an outer concentric spherical cloud in addition, the inner cloud being both denser and more compact since two electrons are involved under attraction by three protons. Figure 2.1 shows the charge cloud concept, which is probably closer to reality, together with the simplicity of illustration of the Rutherford model.

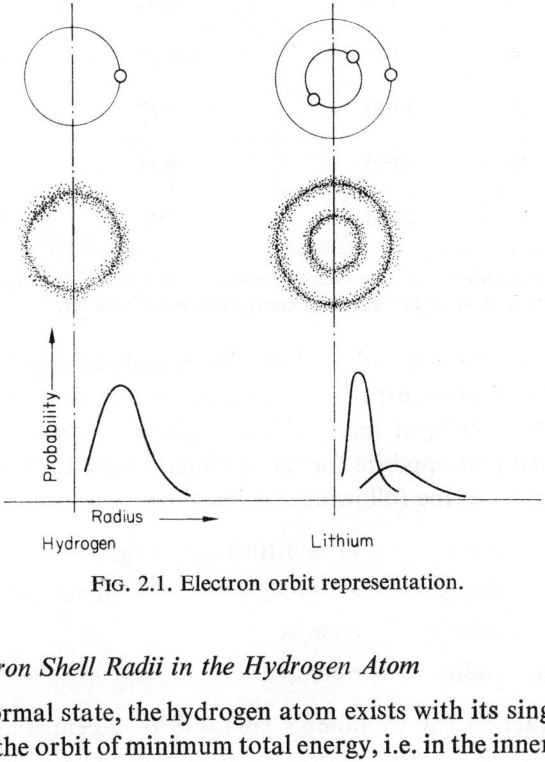

FIG. 2.1. Electron orbit representation.

2.1.2. Electron Shell Radii in the Hydrogen Atom

In the normal state, the hydrogen atom exists with its single electron occupying the orbit of minimum total energy, i.e. in the innermost shell. The addition of discrete quantities of energy in the form of radiation or heat which can be absorbed by the electron leads to successive jumps to orbits in outer shells of progressively higher energy level. The radius of each shell of the hydrogen atom in the excited state can be

TABLE 2.2. Electron Shell Radii and Electron Energies for Hydrogen

Shell	n	n^2	Radius, cm$\times 10^{-8}$	Energy, erg$\times 10^{-12}$	Energy change between adjacent shells, erg$\times 10^{-12}$
K	1	1	0·53	—21·79	
L	2	4	2·12	–5·45	16·34
M	3	9	4·76	—2·42	3·03
N	4	16	8·47	—1·36	1·06
O	5	25	13·23	—0·87	0·49
P	6	36	19·05	—0·61	0·26
Q	7	49	25·93	—0·45	0·16

Note. Negative signs indicate energy *release* during initial location of electron, hence electrons in *K* shell possess *least* energy (derived from ref. 7).

determined by an examination of the forces and the angular momenta involved. The electron experiences an inwards electrical attraction and an outwards centrifugal force which, if circular motion is assumed, must be equal and opposite for the stationary state to be maintained. For an electron of the following specifications;

$$\begin{aligned} \text{mass} &= m = 9{\cdot}1083\times 10^{-28}\ \text{g},\\ \text{charge} &= e = 4{\cdot}8028\times 10^{-10}\ \text{statcoulomb},\\ \text{velocity} &= v\ \text{cm/s},\\ \text{orbiting radius} &= r\ \text{cm}, \end{aligned}$$

inwards attraction (by Coulomb's law) $= e^2/r^2$ = centrifugal force = $= mv^2/r$.

Hence, $$v^2 = e^2/mr. \tag{2.1}$$

By using the arbitrary assumption that the angular momentum of an electron in any stationary state is equal to $(hn/2\pi)$, where n is an integer and h is Planck's constant ($=6{\cdot}625\times10^{-27}$ erg-s), Bohr obtained close agreement between theoretical and experimental results. Hence, angular momentum$=mvr=hn/2\pi$.

Therefore, $$v^2=\frac{n^2h^2}{4\pi^2m^2r^2}=\frac{e^2}{mr},$$

and $$r=\frac{n^2h^2}{4\pi^2me^2}=(0{\cdot}529\times10^{-8})\,n^2\text{ cm}.$$

The integer values of n relate progressively outwards to the main electron shells and are thus the principal quantum numbers, resulting in the shell radii shown in Table 2.2.

2.1.3. Electron Energy

Since, by Coulomb's law, the electrostatic attraction force between the electron and its corresponding nuclear proton equals e^2/r^2, the electrostatic potential energy released in bringing the electron in to the radius r from infinity $=\int_r^\infty \frac{e^2}{r^2}\cdot dr=-e^2/r$.

Due to its motion, the kinetic energy of the electron $=\frac{1}{2}mv^2$. Since $v^2=e^2/mr$, from equation (2.1), the kinetic energy $=e^2/2r$.

Thus, total energy $=-e^2/r+e^2/2r=-e^2/2r=\dfrac{-21{\cdot}79\times10^{-12}}{n^2}$ erg.

Conversely, $+e^2/2r$ units of energy would be required to remove the electron completely from the influence of the nucleus, and this is known as the ionization energy, i.e. the energy required to convert the atom into an ion. Electron energies for the different shells in the hydrogen atom are included in Table 2.2.

Energy is released from or absorbed by atoms in discrete quantities known as quanta, as electrons change their orbit from one shell to another. Hence energy changes of this kind can be expressed as

Energy change quantum $=\Delta E=21{\cdot}79\times10^{-12}(n_1^{-2}-n_2^{-2})$, where n_1 and n_2 relate to the two shells concerned. Energy changes between shells are also included in Table 2.2.

The discovery of this detailed arrangement of electron orbits has been made possible by the fact that moving electrons behave as electromagnetic radiations which can be isolated and identified in a spectroscope. Each energy change, therefore, corresponds to a radiation of wavelength λ given by

$$\frac{1}{\lambda} = \frac{E}{hc} = \frac{21{\cdot}79\times10^5}{6{\cdot}625\times2{\cdot}998}(n_1^{-2}-n_2^{-2}) = R(n_1^{-2}-n_1^{-2}),$$

where R, the Rydberg constant, = 109,600, and c = velocity of light = $2{\cdot}998\times10^{10}$ cm/s. Hence, inwards movement of an electron from shell L ($n_1 = 2$), to shell K ($n_2 = 1$), for example, represents a release of an energy quantum of $16{\cdot}34\times10^{-12}$ erg (see Table 2.2) and results in a radiation of wavelength $1{\cdot}217\times10^{-5}$ cm, or 1217 Angstrom units (Å).

2.1.4. *Electron Distribution in Atoms*

The chemical behaviour of an atom is determined largely by the number and disposition of electrons in the ultimate, or outer, shell and sometimes in the penultimate shell also. Atoms with completely filled outer shells have no chemical activity, and these comprise the inert gases. Atoms with one less than the maximum number of outer electrons (the halogens), and those with a single electron in the outer shell (includes the alkali metals) are particularly active chemically.

Electrons arrange themselves into those shells and orbits in an atom which make the total energy a minimum, the atom then being in the normal or ground state. Consequently, the single electron in the hydrogen atom occupies the innermost shell. In the helium atom, the second electron occupies the innermost shell also, but has opposite spin. This follows from the Pauli exclusion principle which states that no two electrons in any atom can have the same four quantum numbers. In the lithium atom, since the innermost shell is filled, the third electron occupies the next outer shell, and so on with additional electrons.

In the heavier, more complex, atoms with multi-sub-orbits in the

TABLE 2.3. Electron Distribution in First Thirty Elements of Periodic Table

Notes	Atomic No.	Shell / Orbit / Element	K s	L s	L p	M s	M p	M d	N s
	1	H	1						
A	2	He	2						
B	3	Li	Core as He	1					
	4	Be		2					
	5	B		2	1				
	6	C		2	2				
	7	N		2	3				
	8	O		2	4				
C	9	F		2	5				
A	10	Ne		2	6				
B	11	Na	Core as Ne			1			
	12	Mg				2			
	13	Al				2	1		
	14	Si				2	2		
	15	P				2	3		
	16	S				2	4		
C	17	Cl				2	5		
A	18	A				2	6		
B	19	K				2	6		1
	20	Ca				2	6		2
D	21	Sc	Core as Ca					1	2
D	22	Ti						2	2
D	23	V						3	2
D	24	Cr						5	1
D	25	Mn						5	2
D	26	Fe						6	2
D	27	Co						7	2
D	28	Ni						8	2
	29	Cu						10	1
	30	Zn						10	2

Notes:

A = inert gas = atom with outer shell or sub-shell complete.
B = alkali metal = atom with single electron in outer shell.
C = halogen = atom requiring one electron for completion of outer shell or sub-shell.
D = transition element = atom requiring electrons in penultimate shell for completion.

TABLE 2.4A. Electron Distribution and Spin in Isolated Elements

Valency	Atomic No.	Shell / Element \ Orbit	K	L				M		
					P				P	
			s	s	P_x	P_y	P_z	s	P_x	P_y
1	1	H	$\frac{1}{2}$							
1	3	Li	$+\frac{1}{2}, -\frac{1}{2}$	$\frac{1}{2}$						
2	4	Be	$+\frac{1}{2}, -\frac{1}{2}$	$+\frac{1}{2}, -\frac{1}{2}$						
3	5	B	$+\frac{1}{2}, -\frac{1}{2}$	$+\frac{1}{2}, -\frac{1}{2}$	$\frac{1}{2}$					
4	6	C	$+\frac{1}{2}, -\frac{1}{2}$	$+\frac{1}{2}, -\frac{1}{2}$	$\frac{1}{2}$	$\frac{1}{2}$				
3	7	N	$+\frac{1}{2}, -\frac{1}{2}$	$+\frac{1}{2}, -\frac{1}{2}$	$\frac{1}{2}$	$\frac{1}{2}$	$\frac{1}{2}$			
2	8	O	$+\frac{1}{2}, -\frac{1}{2}$	$+\frac{1}{2}, -\frac{1}{2}$	$+\frac{1}{2}, -\frac{1}{2}$	$\frac{1}{2}$	$\frac{1}{2}$			
1	9	F	$+\frac{1}{2}, -\frac{1}{2}$	$+\frac{1}{2}, -\frac{1}{2}$	$+\frac{1}{2}, -\frac{1}{2}$	$+\frac{1}{2}, -\frac{1}{2}$	$\frac{1}{2}$			
	12	Mg	$+\frac{1}{2}, -\frac{1}{2}$	$+\frac{1}{2}, -\frac{1}{2}$	$+\frac{1}{2}, -\frac{1}{2}$	$+\frac{1}{2}, -\frac{1}{2}$	$+\frac{1}{2}, -\frac{1}{2}$	$+\frac{1}{2}, -\frac{1}{2}$		
	13	Al	$+\frac{1}{2}, -\frac{1}{2}$	$+\frac{1}{2}, -\frac{1}{2}$	$+\frac{1}{2}, -\frac{1}{2}$	$+\frac{1}{2}, -\frac{1}{2}$	$+\frac{1}{2}, -\frac{1}{2}$	$+\frac{1}{2}, -\frac{1}{2}$	$\frac{1}{2}$	

TABLE 2.4B. Electron Distribution and Spin in Reacting Elements

2	4	Be	$+\frac{1}{2}, -\frac{1}{2}$	$\frac{1}{2}$	$\frac{1}{2}$					
3	5	B	$+\frac{1}{2}, -\frac{1}{2}$	$\frac{1}{2}$	$\frac{1}{2}$	$\frac{1}{2}$				
4	6	C	$+\frac{1}{2}, -\frac{1}{2}$	$\frac{1}{2}$	$\frac{1}{2}$	$\frac{1}{2}$	$\frac{1}{2}$			

Notes: $\frac{1}{2}$ represents spin in one direction.
$+\frac{1}{2}, -\frac{1}{2}$ represents two opposing spins.
Elements of atomic nos. 6 to 8 illustrate law of maximum multiplicity.

outer shells, the order of orbit filling follows the rule of maximum multiplicity which states that, in a free atom, electrons first occupy the three *p*-sub-orbits singly before any pairing takes place, and similarly with the five *d*- and seven *f*-sub-orbits. Electron distribution for the first thirty elements of the periodic table are shown in Table 2.3, and a more detailed representation, indicating the single and paired electrons, is shown in Table 2.4 for a selected number of light elements of particular interest in this study.

The shorthand form of expressing electron distribution is, for example, Ks^2 for the two electrons in the *s*-orbit of the *K*-shell comprising the helium atom. Hence, the electron pattern for elemental zinc shown in Table 2.3 would be expressed as $Ks^2Ls^2Lp^6Ms^2Mp^6Md^{10}Ns^2$. If necessary, the *x*-, *y*- and *z*-sub-orbits can be shown as p_x, p_y and p_z.

When the atom of beryllium takes part in molecular formation, one of the *Ls* electrons transfers to an *Lp* orbit (see Table 2.4B and Section 2.2.1) and the two sub-orbits become pear-shaped. In a similar way, boron forms three sub-orbits and, in both cases, the sub-orbits are planar and regularly disposed to each other. Carbon forms four orbits by the same process but these are tetrahedrally arranged, with their axes mutually disposed at 109·47°. Since one *s*- and three *p*-orbits are involved, this arrangement is known as a hybrid sp^3 orbital.

Atoms of nitrogen, oxygen and fluorine comprise the same basic tetrahedral structure, with two electrons in one or more of the lobes, whereas in the inert atom of neon the two *Ls* electrons lie in a concentric spherical shell and the six *Lp* electrons are distributed in pairs amongst three dumb-bell-shaped lobes p_x, p_y and p_z centred on the nucleus with mutual perpendicularity.

2.2. Molecular Structure

Following the discussion of the release of energy from a single atom, attention turns to the energy involved in the combination of atoms into molecules, and thus the energy released in molecular dissociation and bond rearrangement. This is dependent upon the manner in which the atoms combine, i.e. on the nature and number of atomic bonds.

As indicated in the previous section, the maximum number of electrons in any orbit is two, and the spins must be opposite. Consequently, an orbit can be filled by two electrons, of opposite spin, which are components of different atoms. Should this arrangement represent a level of energy lower than that of the sum of the energies of the two isolated atoms, the conjoined arrangement will be more stable and a molecular bond will have been formed. This is known as a sigma (σ) bond which, as discussed in the following section, is a single covalent bond made up of two shared electrons, one from each atom, and is typical of the bonding process between carbon and hydrogen, i.e. between a hybrid sp^3 orbital and an s-orbital. The carbon-type four-lobe structure can also behave as a hybrid sp^2 orbital, the one s- and the two of the three p-orbits being mutually disposed at 120° in a single plane, with the remaining p-orbit transverse to the plane and symmetrically disposed in the form of a dumb-bell about the nucleus. When two such atoms combine, a single σ-bond forms between orbits of the two sp^2 hybrids, and an additional bond results between the two transverse p-orbits. This additional bond is known as a π-bond, and is weaker than the σ-bond, making the molecule more liable to reaction. Similarly, the four outer orbits may arrange themselves as an sp hybrid with two transverse p-orbits. Combination now results in a σ-bond together with two π-bonds, representing a still further degree of reactivity and corresponding lack of stability.

2.2.1. Valency

The nature and number of atomic bonds are represented broadly by the valency, or combining power, of the elemental atom. As noted earlier, all material tends towards a condition of stability by the release of energy. Hence, in a single atom, electrons migrate towards the innermost orbits and give up energy in so doing. In a bonding reaction between two or more atoms, this tendency illustrates directly one of the processes of bonding, since electron rearrangement is such that outer orbits become filled, and the atoms simulate inert gases so long as they remain combined.

In the classic example of sodium chloride, the sodium atom introduces its single outer electron into the outer orbit of the chlorine atom, and so fills that particular sub-group of orbits (Fig. 2.2). The penultimate shell of the sodium atom, which is already filled, thus acts effectively as the outer shell. The two atoms now become ions, since one has lost an electron and thus gained a positive charge from the resulting unbalanced proton, whereas the other has gained an electron and hence a negative charge. The chemical properties of the two ions, as represented by the outer electron patterns, resemble those of the inert gases neon and argon. Provided the ions stay combined, they remain chemically inert. Since they are inert, and since the positive and negative charges exert an electrostatic attraction, the ions do stay combined, and the resulting compound, sodium chloride, is stable. This is an example of ionic, or electrovalent, bonding.

Two other major types of bonding are visualized, both of which are effected by means of two electrons instead of one. In covalent bonding

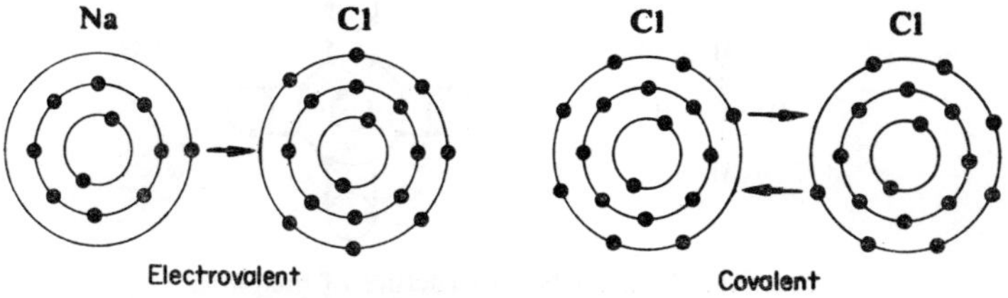

FIG. 2.2. Diagrammatic representation of chemical bonds.

of two atoms, each atom provides one of the two shared electrons, whereas in dative, or coordinate, bonding, one atom provides both electrons.

In any study of fuel chemistry, covalent bonding is a particularly important concept since, for example, the vast range of hydrocarbon materials results mainly from this process. The pattern of covalent bonding can be seen easily from the diatomic molecule example of chlorine included in Fig. 2.2. The attractive force between the two atoms is not so easily apparent, but the concept of resonance gives some appreciation of it. In essence, a number of possible electronic bond patterns is visualized, and structural resonance assumed to take place between them. This resonant bond, which is difficult to represent diagrammatically, has individual characteristics which are influenced by the visualized patterns but are not necessarily an average of them. For example, the energy level of the resonant bond pattern may be lower than those of the visualized patterns, and not the mean of them. Hence, the resonant pattern is the most stable, and the atoms remain combined for this reason. This pattern is described as a resonance hybrid of the visualized patterns, and the energy difference between the hybrid and the most stable (lowest energy level) of the visualized patterns is termed the resonance energy. This is discussed further in connection with heats of formation.

An example of covalent bonding is given in Fig. 2.3 for CH_4, methane, the simplest of the hydrocarbon compounds. The structure is shown first in terms of single bonds, and secondly in terms of the electron arrange-

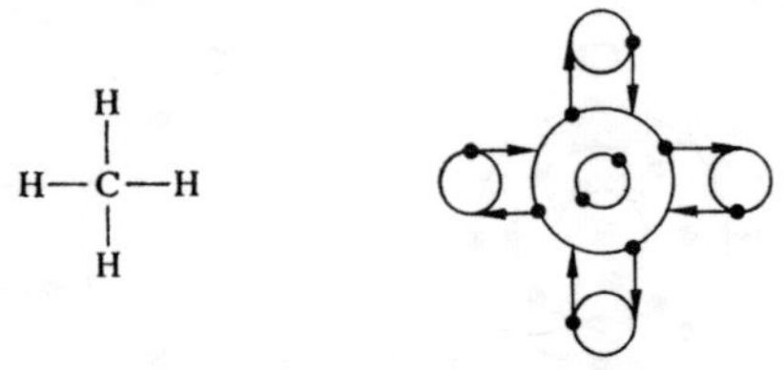

FIG. 2.3. Covalent bond structure of methane.

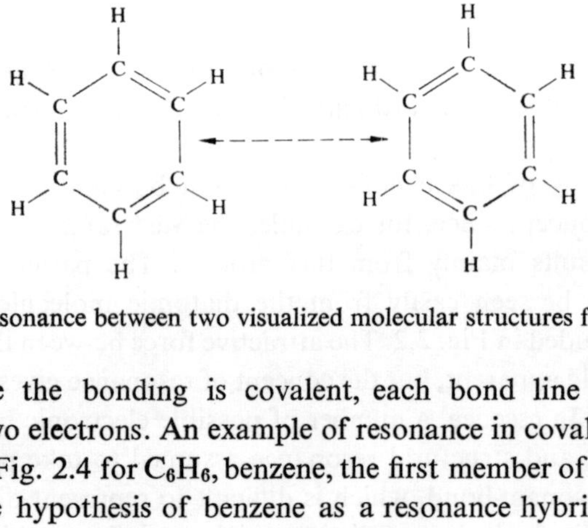

FIG. 2.4. Resonance between two visualized molecular structures for benzene.

ments. Since the bonding is covalent, each bond line of Fig. 2.3 represents two electrons. An example of resonance in covalent bonding is shown in Fig. 2.4 for C_6H_6, benzene, the first member of the aromatic series. The hypothesis of benzene as a resonance hybrid of the two visualized structures shown is strengthened by measurements of atomic spacing within the molecule, as follows;

$$\text{Normal C—C spacing} = 1{\cdot}5\ \text{Å},$$

$$\text{Normal C=C spacing} = 1{\cdot}3\ \text{Å}.$$

The measured carbon-atom spacing of 1·4 Å lies between the two normal values, and is the same for all six bonds within the molecule. The benzene ring structure is envisaged as a linkwork of six σ-bonds between the six sp^2 orbitals, with a double ring of π-bonds resulting from the six transverse p-orbits.

In general, the dative bond occurs with more complex structures such as the ferrocyanides, but the nitrogen atom in the ammonia molecule represents a simple example of a filled outer orbit containing a lone pair of electrons available for further bonding. On the other

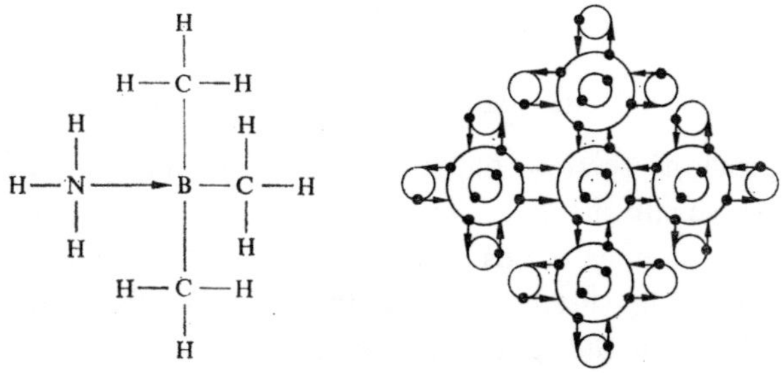

FIG. 2.5. Dative bond between ammonia and tri-methyl borane.

hand, the atom of boron in the molecule of trimethyl borane represents an unfilled outer orbit in which all electrons are utilized in bonding. A bond between these two molecules would be a dative type, with nitrogen as the donor and boron the acceptor, as shown in Fig. 2.5. Although these three major bond types are discussed here as if distinct, actual bonds are considered to be intermediate in nature, possessing some mixed characteristics of each.

From the general considerations outlined above, the numerical value of the valency of an elemental atom is related closely to either the number of outer electrons present, or the number of electrons required to fill the outer shell or an outer sub-orbit. Consequently, some elements exhibit more than one value of valency. The main values of valency for the elements of interest are included in Table 2.4. Of the elements shown in the second period, the values of valency are equal to the number of unpaired electrons in all cases except those of beryllium, boron and carbon. It is assumed, therefore, that values of valency are equal to the number of unpaired electrons in an outer shell, or in a sub-orbit within it, and that in each of the three cases mentioned, the reaction energy released is sufficient to uncouple the paired electrons in the *s*-orbit of the *L*-shell and to divert one of them into a vacant *p*-sub-orbit (Table 2.4B).

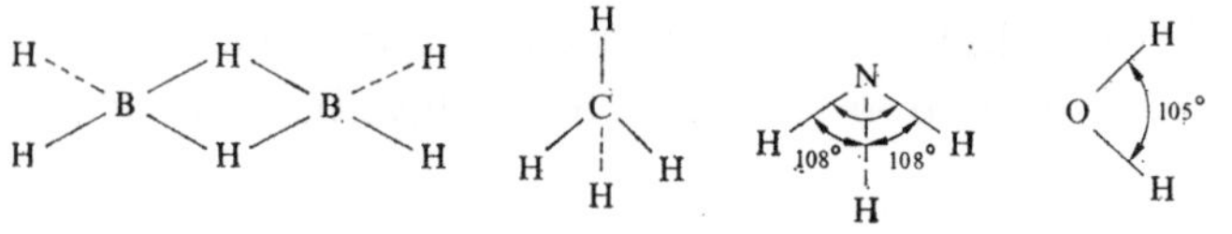

FIG. 2.6. Bond angularity and three-dimensional structure of simple molecules based on H, B, C, N and O. (Dotted line indicates bond inclined to plane of other bonds.)

In the previous section, the mutual perpendicularity was noted between the three sub-orbits comprising the *p*-group in the *L*-shell. Consequently, the valence bonds would be expected to be similarly orientated in space. Measurements show that angular dispositions between the bonds do occur but that, with covalent bonds, the angles are generally in excess of 90°. This follows because the bonds have some ionic character, the resulting like charges exerting a repulsion and forcing the angle open. Molecules based on boron, carbon, nitrogen and oxy-

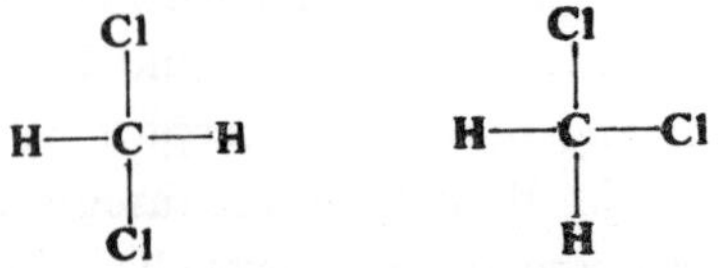

FIG. 2.7. Different two-dimensional structural concepts for CH_2Cl_2.

gen, therefore, would be expected to exhibit angular dispositions between the bonds and, in most cases, a three-dimensional structure (Fig. 2.6). This follows since all four atoms, *when combining with other atoms*, contain at least two unpaired atoms in the mutually perpendicular *p*-orbits.

The concept of a three-dimensional structure for CH_4, for example, is strengthened by the fact that the substitution of two chlorine atoms for two of the hydrogen atoms produces methylene dichloride, CH_2Cl_2, with properties that are unique. From consideration of a two-dimensional structure, however, two materials would be expected, with differing properties (Fig. 2.7). For illustrative convenience, however, the two-dimensional presentation is retained.

2.2.2. *Chemical Energy*

Heat is evolved from a wide variety of chemical reactions including formation, decomposition, oxidation, dilution and neutralization. One of the simplest exothermic chemical reactions available is the decomposition of an endothermic material, i.e. a material which absorbs heat during the process of formation from its component elements. Decomposition triggered off by physical and/or chemical means leads to the release of the absorbed formation heat. By definition, a positive sign indicates that the heat energy has been added during the reaction. Hence, the exothermic heat energies of propulsion interest will be represented by negative quantities. The property requirement for a high-performance fuel of the decomposition type, therefore, can be expressed as a high negative heat of decomposition, and typical values are given in the following Chapter.

As a consequence of the abundance of free oxygen in the Earth's atmosphere, oxidation represents the most common example of the combination reaction, and the most attractive economically. Energy storage fluids, therefore, normally comprise a fuel and oxidant pair, and this arrangement is so commonplace that the terms "fuel" and "oxidant" are retained in those cases where alternative reactant materials are used, e.g. fluorine as an "oxidant" with hydrogen as a fuel. Since both fuels and oxidants may contain proportions of such elements as carbon, hydrogen, oxygen and nitrogen, and since oxygen may not appear in either, the usage of these terms is evidently somewhat loose, and the concept of the reactant pair is more accurate.

The proportions of materials involved in chemical reactions are dependent upon the disposition of the outer electrons around the elemental nuclei, as represented by the valency. The energy released from such reactions is a function of the energy absorbed in the formation of both the reactants and their products, as represented by the heats of formation. The heat of reaction can be calculated from a knowledge of the heats of formation of the reactants and products, and of the proportions involved. The following discussion indicates how these quantities provide a measure of the property requirements for high-energy output and high thrust. Reactions at constant pressure are assumed throughout.

2.2.3. *Heats of Formation and Reaction*

In general, the heat energy change accompanying any chemical reaction may be described as the heat of reaction and represented by ΔH_r, but alternative nomenclature is used with specific reactions such as the heat of formation ΔH_f, the heat of atomization ΔH_a, and the dissociation bond energy $D(X\text{—}Y)$. The understanding of reaction processes provided by the study of chemical kinetics is based upon the heat of formation which is defined as the reaction heat absorbed or released when one mole† of a compound is formed from its elements. Since heat changes only are measurable, a datum condition, usually 25° C (298° K) and 1 atmosphere, is employed, represented by the superscript °. The elemental molecules are considered to exist initially at the datum condition in their most stable state, and the resulting compound is brought back to this condition by the addition of ΔH_f°, i.e.

Elemental molecules →	Compound molecules	⟶	Compound molecules
(298° K & 1 atm)	(T° K & 1 atm)	ΔH_f° added or subtracted	(298° K & 1 atm)

(2.2)

or, more simply,

Elemental molecules	⟶	Compound molecules
(298° K & 1 atm)	ΔH_f° added or subtracted	(298° K & 1 atm)

† The mole unit is normally employed in thermochemistry, whereas unit mass is the more significant basis in aeronautics and astronautics. The main thermochemical arguments are thus used here on a mole basis, and the conclusions converted to a unit mass basis as required.

Oxidation reactions can be treated in a similar manner, and the heat of oxidation reaction defined as the heat absorbed or released when one mole of the material reacts completely with oxygen, the reactants and products being at the datum conditions. The heat of reaction, which will usually be exothermic, may be illustrated as follows;

$$\underset{(298^\circ\text{ K \& 1 atm})}{\text{Reactants}} \xrightarrow{\quad \Delta H_r^\circ \text{ released} \quad} \underset{(298^\circ\text{ K \& 1 atm})}{\text{Product}}$$

and further defined as the heat released by the products of a completely oxidized mole of material in order to regain the datum conditions. The heat of reaction, therefore, is equal to the difference between the heats of formation of the products and of the reactants, since

Reactant elements → (ΔH_f° reactants) → Reactant compounds → (ΔH_r°) → Product compounds

Reactant elements → (ΔH_f° products) → Product compounds

i.e. $\Delta H_r^\circ = \Delta H_f^\circ$ products $- \Delta H_f^\circ$ reactants (2.3)

The property requirements of combination-type reactants may thus be expressed as a high negative heat of reaction. This implies reactants of high positive (or low negative) heat of formation, and products of high negative heat of formation. These points are illustrated in the upper portion of Fig. 2.8.

The formation process, in fact, comprises two stages since the elemental molecules must first be converted into free gaseous atoms before rearrangement is possible to the resultant compound molecule, i.e. "heat of atomization" must be supplied to break the existing bonds prior to the formation of new bonds and the release of corresponding "bond energy". In general terms, this would appear as follows:

Elemental molecules ⟶ Gaseous Atoms ⟶ Compound molecule

Elemental molecules ⟶ Gaseous Atoms	Gaseous Atoms ⟶ Compound molecule
Energy added for bond disruption (atomization heat)	Energy released from bond formation (bond energy)

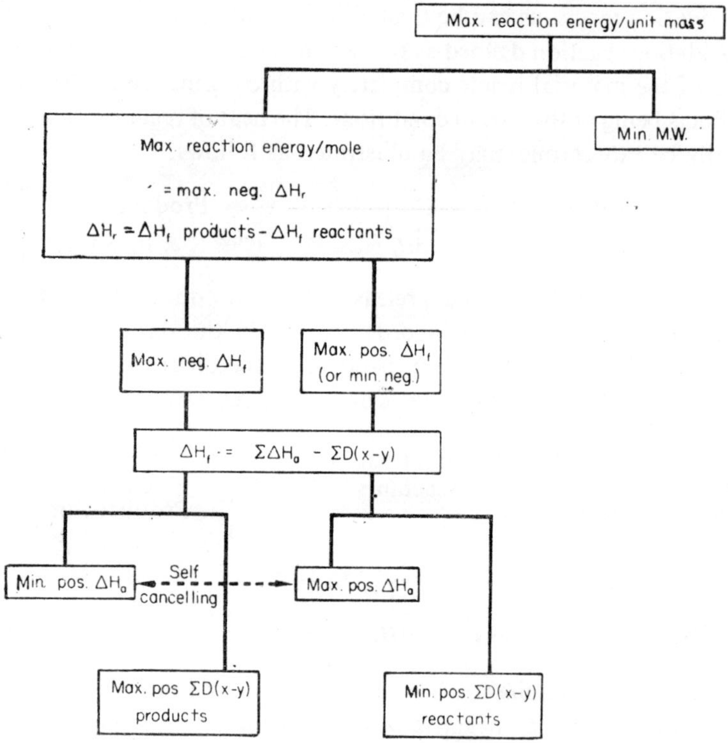

FIG. 2.8. Reaction energy diagram.

As shown above, atomization heat is represented by ΔH_a, and bond energy by $D(X—Y)$ to indicate the energy required to dissociate an $X—Y$ bond. The resulting heat of formation, therefore, is the difference between these two quantities, i.e.

$$\Delta H_f^\circ = \sum \Delta H_a^\circ - \sum D(X—Y).$$

Typical values of these quantities are given in Chapter 3.

The property requirements for combination-type reactants given above may now be expanded a stage further and expressed as high positive atomization heat and low positive bond energy, giving products of low

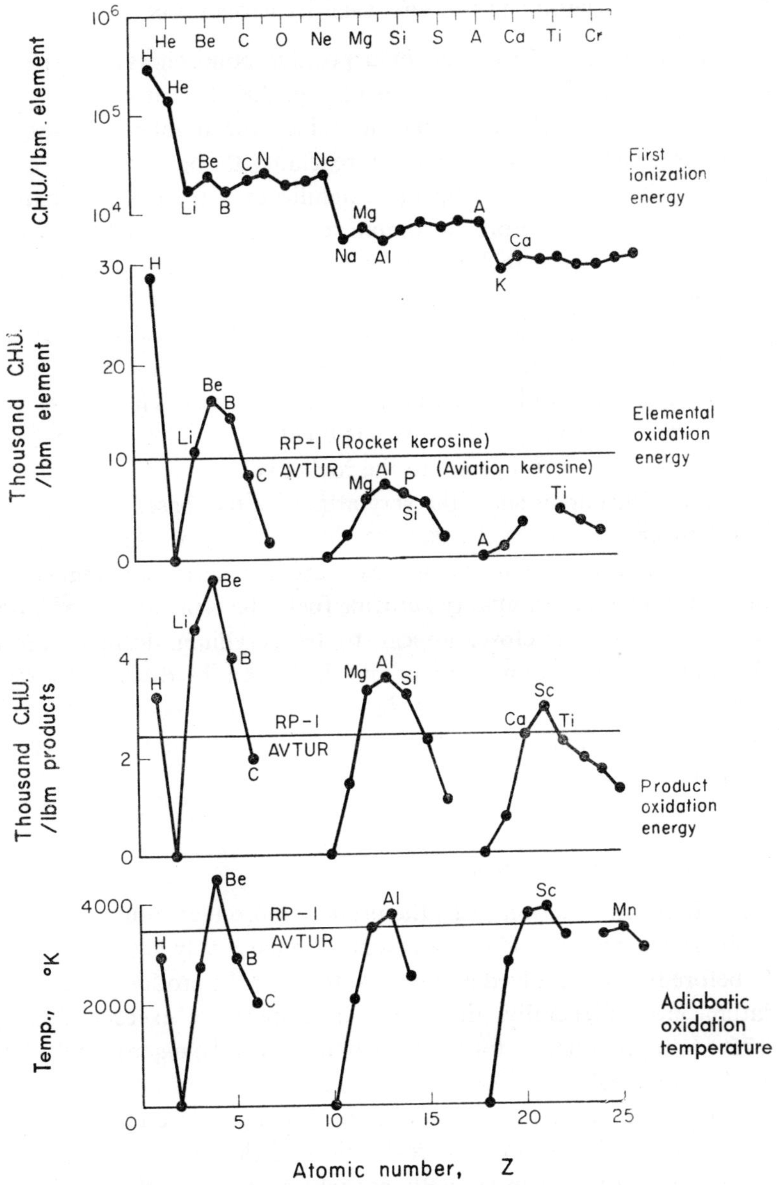

FIG. 2.9. Elemental oxidation energies and temperatures (derived from refs. 19,21 and 23).

positive atomization heat and high positive bond energy. This point is illustrated in the lower portion of Fig. 2.8. The atomization heat requirements are self-cancelling and, since the number of moles of products generally exceeds that of the reactants, the bond energy requirements of the products are the most significant of these four quantities. In general terms, therefore, maximum reaction energy would be expected from reactants with low-energy bonds which form products with high-energy bonds, i.e. unstable reactants forming stable products.

The oxidation heating values of the elements have been determined and are shown in Fig. 2.9. This composite figure presents these values on unit mass bases of both elements and products. The former is of direct use for ramjet and other air-breathing applications, whereas the latter is more appropriate to the rocket vehicle carrying all its own reactants. The curves show the very attractive fuel-based heating value of hydrogen, and product-based heating value of beryllium. Included in the figure for reference purposes are the corresponding heating values for an aviation-quality kerosine fuel. The elements of oxidation-energy interest, therefore, appear to be beryllium, lithium, boron, aluminium, magnesium, silicon and hydrogen in descending order of attraction. A wider variety of high-energy fuels is available, however, since the compounding of these and other elements can result in further attractive values of heats of formation and reaction.

2.2.4. Dissociation and Equilibrium

As shown in equation (2.2), the product molecules resulting from an exothermic reaction tend towards some comparatively high temperature T° before the heat evolved is extracted to bring the products back to the datum level. Generally, this reaction temperature exceeds the level (1750° C approximately for hydrocarbon fuels and oxygen) at which the product molecules become unstable and begin to dissociate back into their reactant form. This lowers the peak temperature reached during combustion, and permits full realization of the reaction energy only if the temperature is subsequently brought down below the dissociation level.

In general terms, a dissociating reaction could be represented as follows:

$$A+B+C \longleftrightarrow D+E,$$

where the sign $\leftrightarrow$ indicates that the reaction is reversible. This expression illustrates the *basic process* of the reaction, and shows which species are involved, together with their initial and final molar quantities.

After the reactions have proceeded for a period of time, the instantaneous condition could be represented as follows:

$$aA+bB+cC \rightleftharpoons dD+eE, \tag{2.4}$$

where the sign $\rightleftharpoons$ indicates, again, that the reaction is reversible, but that a, b, etc., are the moles of materials coexisting in the system at the given instant, i.e. that the sign has the sense of a plus sign. If, for example, one mole each of A, B and C are the initial materials, the reaction has proceeded as follows;

$$A+B+C \longrightarrow aA+bB+cC+dD+eE$$

Initial condition	Instantaneous condition

The proportions of materials in coexistence under given conditions are predictable from the rates of the forward and reverse reactions, which are derived from chemical kinetics. As stated in the Law of Mass Action, first enunciated by Guldberg and Waage in 1863: "The rate of a chemical reaction is proportional to the products of the active concentrations of the reactants." Concentrations are expressed normally in moles per unit volume of system, hence, in equation (2.4),

Rate of rightwards reaction $\propto abc/V^3$,

and Rate of leftwards reaction $\propto de/V^2$,

where V is the total volume of the system.

Since molar concentration is proportional to partial pressure, and chemical reaction rates are also sensitive to temperature, the two reaction rates shown above may be expressed as follows:

Rate of rightwards reaction $= k_R f_R(T)\, p_A p_B p_C$,

and Rate of leftwards reaction $= k_L f_L(T)\, p_D p_E$,

where k_R and k_L are rate constants. In the equilibrium condition, the two rates are equal, hence,

$$\frac{p_D p_E}{p_A p_B p_C} = \frac{k_R f_R(T)}{k_L f_L(T)} = K = \text{Equilibrium constant,} \qquad (2.5)$$

which is itself temperature dependent.

By convention, products appear on the right-hand side of the reaction expression, and as the numerator in the equilibrium constant expression. Strictly, since the reaction is reversible, products and reactants cannot be precisely distinguished. In general, the value of K is dependent upon partial pressure units, and atmospheres are commonly employed.

If the initial proportions of materials do not match those in the basic process expression, the rates of reaction and hence the equilibrium condition will be changed. For example, if the concentration of material B is trebled, with the same basic process, the initial condition becomes $A+3B+C$, hence the initial reaction rate is also trebled. Further, if the concentration of C is then also trebled, the initial reaction is increased by a factor of 9. In the special case where B and C are the same material, the basic process may be expressed as

$$A+2B \longleftrightarrow D+E,$$

and the reaction rate is seen to be proportional to the square of the concentration of B. In terms of partial pressures, therefore;

$$K = \frac{p_D p_E}{p_A (p_B)^2}.$$

Hence, the general expression for the equilibrium constant in a reversible reaction,

$$aA+bB+\ldots \rightleftharpoons xX+yY+\ldots$$

controlled by the basic process,

$$n_A A+n_B B+\ldots \longleftrightarrow n_X X+n_Y Y+\ldots$$

is given by:

$$K = \frac{(p_X)^{n_X}(p_Y)^{n_Y}\ldots}{(p_A)^{n_A}(p_B)^{n_B}\ldots}$$

where n_A, n_B, etc., are the numbers of moles of A, B etc., in the basic process expression. In each case, the partial pressure of a material is related to both the number of moles of the material (a, b, etc.) and the total number of moles (N) present in the equilibrium expression, and to the total pressure (P), e.g. $P_X = (x/N)P$.

In the case of oxidation reactions with hydrogen to H_2O, and with carbon through CO to CO_2, the basic processes of the final phases of the reactions are:

$$H_2 + \tfrac{1}{2}O_2 \longleftrightarrow H_2O \tag{2.6}$$

and

$$CO + \tfrac{1}{2}O_2 \longleftrightarrow CO_2. \tag{2.7}$$

At any equilibrium condition represented by

$$aH_2 + bO_2 \rightleftharpoons cH_2O, \quad \text{and} \quad dCO + eO_2 \rightleftharpoons fCO_2,$$

application of the equilibrium constants

$$K_{H_2O} = \frac{P_{H_2O}}{P_{H_2}(P_{O_2})^{1/2}}, \quad \text{and} \quad K_{CO_2} = \frac{P_{CO_2}}{P_{CO}(P_{O_2})^{1/2}}$$

permits the derivation of the equilibrium proportions of each material present.

In the oxidation of hydrocarbons, the half mole of oxygen produced by dissociation of the mole of CO_2 in equation (2.7) may be considered as the oxygen required to oxidize the mole of H_2 in equation (2.6). These two equations, therefore, may be combined to give a basic process,

$$CO + H_2O \longleftrightarrow CO_2 + H_2,$$

which is the Water–Gas Reaction, and which leads to the Water–Gas Reaction Equilibrium Constant:

$$K_{WG} = \frac{P_{CO_2}P_{H_2}}{P_{CO}P_{H_2O}}.$$

An assumed value of combustion temperature permits the use of published values of equilibrium constants to derive the relative quantities of reactant and product in coexistence. This in turn permits calculation of quantities of energy released by the burning of part of the reactants, and absorbed by the products formed, an energy balance being achieved by an iterative process to give the true combustion temperature for the given conditions. Comparison of the actual energy release thus determined with the theoretical heat of reaction for complete combustion indicates the loss of energy resulting from dissociation, and provides a more realistic performance figure for the reactants under consideration. At the very high temperatures involved with reactions with pure oxygen, dissociation of product molecules becomes extensive and complex, involving free atoms and radicals, coupled with ionization of atoms.

The full realization of the energy potential of the reactants is possible only if the conversion of energy to thrust brings the propellant temperature down below the ionization and dissociation levels before the propellant is ejected from the propulsion system. The extent of this realization is represented by the frozen-flow efficiency of Chapter 1.

2.3. Nuclear Structure

In general, an atomic nucleus comprises a number of positively-charged particles together with a number of neutral particles of slightly greater mass. These particles are known respectively as protons and neutrons, and collectively as nucleons. Several other kinds of particle have been postulated, and their presence detected by experiment. One of the remarkable phenomena noted in nuclear physics is the apparent facility for interchange between the various kinds of particle.

Each nuclear species capable of more than transient existence is termed a nuclide, the simplest of which is represented by hydrogen with a nucleus consisting of one proton only. Uranium, on the other hand, represents one of the most complex stable nuclei known, with 92 protons and 146 neutrons. The following symbols are adopted to define the numbers and masses of the particles;

Z = atomic number = number of protons,
M = atomic mass = true mass of atom in atomic mass units (1 amu = $1{\cdot}6603 \times 10^{-24}$ g),

A = mass number = number of nucleons = M rounded off to an integer.

It follows that N, the number of neutrons, = $(A-Z)$.

Alternative representations commonly used for a particular nuclide are as follows:

$$ {}^{A}_{Z}\mathsf{X}, \quad {}_{Z}\mathsf{X}^{A}, \quad \text{or} \quad \mathsf{X}^{A}_{Z}, $$

where X is the chemical symbol for the element. Hence hydrogen, with one proton only, is more completely represented as, say, H^1_1, and helium, with two protons and two neutrons, as He^4_2. The form of uranium mentioned above would be written as U^{238}_{92}.

The diameters of protons, neutrons and electrons are not clearly defined but are all of order 3×10^{-13} cm. The masses of protons and neutrons differ slightly and are much greater than that of the electron. Protons and electrons attract each other by means of an electrostatic force through their equal and opposite electrical charges, whereas protons and neutrons attract each other by means of a "nucleic charge" force. These factors are summarized in Table 2.5.

TABLE 2.5. Details of Nucleons and Electrons

Particle		Proton	Neutron	Electron
Mass	gram (approx.)	$1{\cdot}66\times10^{-24}$	$1{\cdot}66\times10^{-24}$	$9{\cdot}1\times10^{-28}$
	amu	1·00758	1·00897	0·00054
	rel. electron	1836	1838	1
Electrostatic charge	coulomb	$1{\cdot}6\times10^{-19}$	0	$-1{\cdot}6\times10^{-19}$
	rel. proton	1	0	−1
Nucleic "charge"	rel. proton	1	1	10^{-8}

In general, the diameter of the nucleus is given by the equation

$$ d = 3\times10^{-13}A^{\frac{1}{3}} \text{ cm}, $$

which gives nuclear diameters ranging from 3×10^{-13} cm for H^1_1 to 18×10^{-13} cm for U^{238}_{92}. In view of the direct proportionality with A, the number of nucleons, it follows that the degree of packing of the nucleons is the same for all nuclei. Hence, the "nucleic charge" forces arise from the close proximity of adjacent nucleons and not from some central nucleic force. This concept of the nucleus is termed the liquid drop model since the nucleons correspond to molecules of liquid in a drop. It is also found that particular stability exists with such "magic numbers" of protons or neutrons as 2, 8, 14, 20, 28, 50, 86 and 126 which suggests that nucleons arrange themselves into series of shells in a manner comparable with the electron orbits around the nucleus.

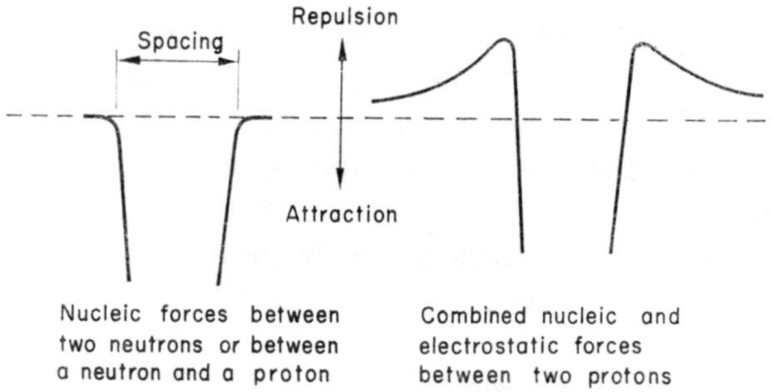

FIG. 2.10. Nucleonic forces.

The variation of nucleic forces with nucleon spacing may be represented as in Fig. 2.10. Corresponding curves obtain for the energy levels of the nucleons, energies resulting from attraction being shown as negative since potential energy is lost by the nucleon when moving under the action of an attractive force. This fact, and the shape of the curve, give rise to the term "potential energy well", as described in Section 1.4. The addition of electromagnetic energy to the nucleons raises the level of the bottom of the well until eventually the brink is reached and the nucleus disintegrates.

A tentative mechanism of nucleic force action is provided by the concept of the meson, which is a particle of mass ranging from about 206 to 970 times that of an electron, and possessing a positive, negative, or zero electrostatic charge. A neutron and proton, therefore, would be expected to influence each other, i.e. to attract, by a continual exchange of a negatively-charged meson, the mass of the meson reducing its effectiveness rapidly to zero on reaching a certain spacing between the nucleons. The meson does not represent additional mass to the nucleus since it is contained within and forms part of the nucleons themselves.

Nuclei of the elements of the periodic table are built up progressively by the addition of a proton at each step, and the number of protons determines the chemical identity of the resulting atom. The presence of two neutrons with the two protons was noted earlier in the case of the second lightest element, helium. Neutrons also are added progressively to elemental nuclei and in slightly greater proportion, so that those nuclei showing the greatest stability exhibit, in general, a slight excess of neutrons over protons. In most elements, the nuclei can still exist stably with a small variation in the number of neutrons within them. These alternative forms of an element are known as isotopes, and the nuclear representation is as follows:

$$X_Z^{A1}, \quad X_Z^{A2}, \quad \text{etc.}$$

Since the number of protons, and hence the number of electrons orbiting in the atom, are the same throughout the group, the isotopes differ in mass but are identical chemically, and cannot be separated by chemical means. Different elements with a common number of neutrons are termed isotones, while those with a common number of nucleons are termed isobars and may be represented as follows:

$$X_{Z1}^{A}, \quad Y_{Z2}^{A}, \quad \text{etc.}$$

Since Z, the number of protons, gives the same information as the chemical symbol, nuclear representation is often simplified to

$$X^A, \text{ i.e. (chemical symbol)}^{\text{number of nucleons}}.$$

A plot of the number of neutrons, N, against the number of protons,

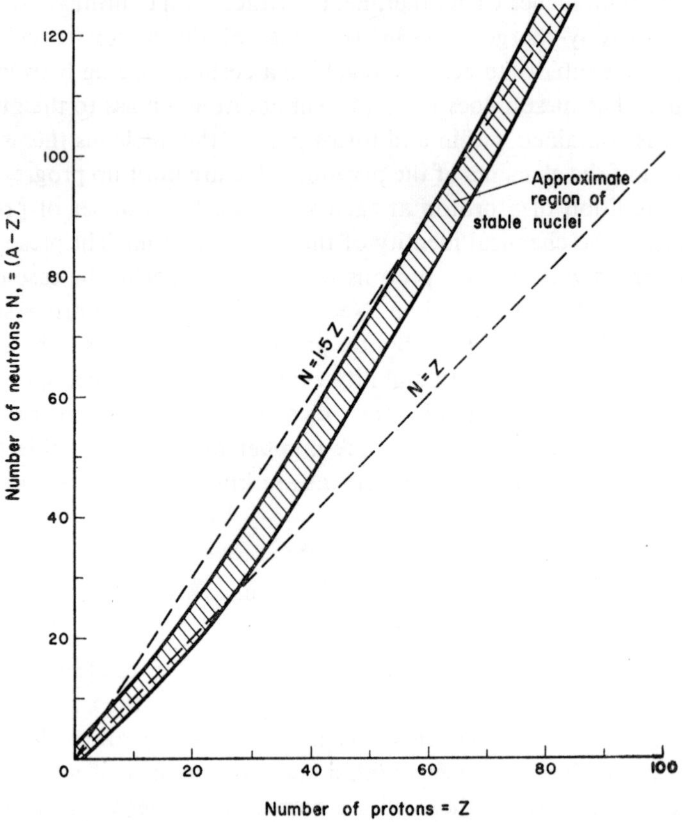

FIG. 2.11. Numbers of neutrons and protons in stable nuclei.

Z, as in Fig. 2.11, confirms that stability with the heavier nuclides is favoured by an increased neutron/proton ratio. All known stable isotopes lie within the shaded area in the figure, any nucleus finding itself outside this area returning to it by suitable rearrangement and emission of nucleic particles. A nucleus located above the shaded area, for example, could achieve stability by means of a neutron emission. The reduced neutron/proton ratio of the nucleus would bring its location on the figure vertically downwards towards the stable area. Alternatively, the neutron may transform into, and emit in the form of, a proton and an electron.

2.3.1. Nuclear Energy

In building an atomic nucleus from protons and neutrons, as described above, a remarkable phenomenon emerges in connection with the total and individual masses. In the simplest example of helium, He_2^4, summation of the particle masses gives;

Two protons at 1·00758 amu each	= 2·01516 amu
Two neutrons at 1·00897 amu each	= 2·01794 amu
Total by summation	= 4·03310 amu
But measured nucleic mass	= 4·00280 amu

The difference represents a mass defect of 0·03030 amu per nucleus, or 0·00758 amu per nucleon. This mass was lost in some way during the process of formation of the nucleus from its components. This violates one of the laws of classical mechanics relating to the conservation of mass, and leads to revised laws to suit the atomic scale of dimensions which concern the conservation of (mass+energy), and consequently the equivalence of mass and energy. The rate of exchange is a function of the speed of light, i.e.

$$(E+mc^2) = \text{constant},$$

and

$$\Delta E = c^2\,\Delta m.$$

The value of c is approximately 3×10^{10} cm/s, and the unit of energy commonly employed is the electron volt, denoted by eV, which is the energy of an electron falling through a potential difference of 1 volt. For larger quantities of energy, the more convenient unit is the million- or mega-electron volt, denoted by MeV. The equivalence between units of mass and energy is shown in Table 2.6.

TABLE 2.6. Equivalence of Mass and Energy

Mass	Energy			
	Kwh	CHU	J	MeV
1 lbm	113×10^{8}	$21{\cdot}5 \times 10^{12}$	$40{\cdot}8 \times 10^{15}$	225×10^{27}
1 g	25×10^{6}	$47{\cdot}4 \times 10^{9}$	9×10^{13}	$56{\cdot}2 \times 10^{25}$
1 amu	$41{\cdot}5 \times 10^{-17}$	$78{\cdot}7 \times 10^{-15}$	$14{\cdot}9 \times 10^{-11}$	931

The mass defect of a nucleus, therefore, represents a quantity of energy emitted during the synthesis process. This energy is termed the binding energy of the nucleus, and is equivalent to the supply of energy required to disintegrate the nucleus into its free nucleons. The binding energy, B, of the helium nucleus discussed above is obtained by applying

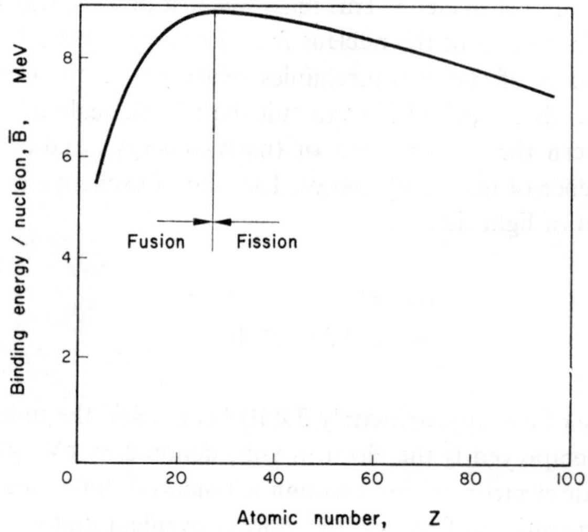

FIG. 2.12. Approximate variation of binding energy per nucleon with atomic number.

the mass/energy relationship to the mass defect, i.e.

$$B \text{ per nucleus} = 0{\cdot}03030 \times 931 = 28{\cdot}2 \text{ MeV},$$

and $$\bar{B} \text{ per nucleon} = 28{\cdot}2/4 = 7{\cdot}05 \text{ MeV}.$$

The variation of $\bar{B}$ throughout the periodic range can be expressed as:

$$\bar{B} = (1.00813\, Z + 1.00897\, N - M)\, 921/A \text{ MeV}, \tag{2.8}$$

where the figure 1·00813 represents the combined mass of each proton and its corresponding planetary electron. This variation is shown in Fig. 2.12. The curve reaches a peak in the region of Fe^{56}, and indicates that energy is released as the peak is approached from either side, i.e. in both the fusion of light nuclei and the fission of heavy nuclei, the former providing the greater range of energy release.

Symbols

A = Mass number
B = Binding energy
$\bar{B}$ = Binding energy per nucleon
c = Velocity of light
$D(X-Y)$ = Bond energy to dissociate X and Y at datum conditions
d = Diameter of nucleus
E = Energy
e = Charge of electron
eV = Electron volt
$f_R(T)$ = Function of temperature in rightwards reaction
ΔH_a° = Heat of atomization at datum conditions
ΔH_f° = Heat of formation at datum conditions
ΔH_r° = Heat of reaction at datum conditions
h = Planck's constant
K = Equilibrium constant
k_R = Rate constant in rightwards reaction
M = Atomic mass
m = Mass
N = Number of neutrons

n = An integer
P_x = Partial pressure of material x
R = Rydberg constant
r = Orbit radius of electron
V = Total volume of reacting system
v = Velocity of electron
Z = Atomic number
λ = Wavelength of radiation

CHAPTER 3

CHEMICAL PROPULSION

PROPULSION by the chemical rocket imposes two different duties upon the reactants; the release of energy, and the production of propellant materials. The reaction energy requirement is seen to be dependent upon the heats of formation of the reactants and their products, and thus upon their bond energies (Chapter 2). The required properties of the reaction products which constitute the propellants centre mainly on the reaction temperature, mean molecular weight, and mean value of γ (Chapter 1).

With endothermic reactants, energy is released directly by decomposition either partially to component compounds or completely to component elements. The high negative heat of decomposition required is realized by the high positive heat of formation. Materials of this kind are shown in Table 3.1 but none of these has yet found application as

TABLE 3.1. Decomposition Energy Release of Endothermic Compounds

Compound	Formula	M.W.	Decomposition energy release	
			kcal/mole	cal/g = CHU/lbm
Acetylene	C_2H_2	26	54·9	2110
Ammonium azide	NH_4N_3	60	96·2	1604
Hydrazoic acid	HN_3	43	61·0	1420
Dicyanogen	C_2N_2	52	65·5	1260
Hydrocyanic acid	HCN	27	23·6	875
Ozone	O_3	48	34·5	720

Representative values derived from various sources.
1 CHU/lbm = 4186·8 J/kg.

a monopropellant due to explosion and toxicity hazards. Less energetic materials, such as concentrated hydrogen peroxide (see later) and hydrazine, are used in small propulsion units for attitude control, station keeping, etc., and as an energy source for propellant pumping.

The heat of formation is zero for all elements in their standard state at the datum condition since the heats of atomization self-cancel with the bond energies, thus $\Delta H_f^\circ = 0$ for $H_2(g)$, $O_2(g)$, $N_2(g)$, etc., where symbols (g), (l) and (s) are used to indicate gaseous, liquid and solid states respectively. Carbon exists in a number of solid allotropic forms, hence graphite, the form of interest for propulsive purposes, is represented as C(gr), for which ΔH_f° also equals zero.

As an example, the formation of gaseous hydrogen peroxide from gaseous hydrogen and oxygen would appear as follows:

$$H_2(g)+O_2(g) \xrightarrow[\Delta H_f^\circ = -31\cdot 8 \text{ kcal/mole}]{} H_2O_2(g)$$

and is an exothermic reaction. The heat of formation of hydrogen peroxide in the liquid state, which is the normal state at the datum conditions, is thus found by adding the exothermic latent heat of condensation at the datum conditions, i.e.

$$\begin{aligned} \Delta H_f^\circ\, H_2O_2(l) &= \Delta H_f^\circ\, H_2O_2(g) + \Delta H_{\text{cond}}^\circ\, H_2O_2(g) \\ &= -31\cdot 8 - 13\cdot 0 \\ &= -44\cdot 8 \text{ kcal/mole.} \end{aligned}$$

Typical values of ΔH_f° are listed in Table 3.2, arranged in descending order of propulsion attraction.

Hydrogen peroxide, in both gaseous and liquid states, is an exothermic formation compound, and is thus endothermic on decomposition into its elements, i.e.

$$H_2O_2(l) \longrightarrow H_2(g)+O_2(g)$$

$$\Delta H_r^\circ = +44\cdot 8 \text{ kcal/mole}$$

However, in the presence of a catalyst, hydrogen peroxide decomposes into water and oxygen in an exothermic manner, as follows:

$$H_2O_2(l) \longrightarrow H_2O(l)+\tfrac{1}{2}\,O_2(g),$$

$$\Delta H_r^\circ = -23\cdot 5 \text{ kcal/mole}$$

TABLE 3.2. Heat of Formation, ΔH_f°

		Material		M.W.	kcal/mole	CHU/lbm
Reactants	Fuels	Acetylene	$C_2H_2(g)$	26	54·9	2110
		Hydrazine	$N_2H_4(l)$	32	12·1	380
		Benzene	$C_6H_6(l)$	78	11·7	150
		Hydrogen	$H_2(g)$	2	0	0
		Ammonia	$NH_3(g)$	17	-11·0	- 650
		Ethane	$C_2H_6(g)$	30	-20·2	- 680
		Methane	$CH_4(g)$	16	-17·9	—1120
		Diborane	$B_2H_6(g)$	28	—33·0	—1180
		Pentaborane	$B_5H_9(l)$	64	—87·0	—1360
		Methanol	$CH_3OH(l)$	32	—57·0	-1790
	Oxidants	Ozone	$O_3(g)$	48	34·0	710
		Oxygen	$O_2(g)$	32	0	0
		Nitric acid	$HNO_3(l)$	63	—41·4	— 660
		Hydrogen peroxide	$H_2O_2(l)$	34	—44·8	—1320
Products		Water	$H_2O(l)$	18	—68·3	—3800
		Carbon dioxide	$CO_2(g)$	44	—94·1	—2140
		Carbon monoxide	$CO(g)$	28	—26·4	— 950
		Nitrogen	$N_2(g)$	28	0	0

Representative values derived from ref. 17 and other sources.
1 CHU/lbm = 4186·8 J/kg.

and it is this reaction which makes hydrogen peroxide attractive as an energy source and a monopropellant. In practice, additional energy is obtainable by subsequent combustion of a fuel with the oxygen released by decomposition, thus converting from a monopropellant to a bipropellant system.

Relative values of ΔH_f for reactants and products must be treated with caution in view of the weighting necessary to account for the molecular proportions. Generally, a single mole of fuel produces more than one mole of products (in the example below, one mole of propane produces seven moles of oxides) and the heats of formation of the products tend to be more significant. In comparing two alternative reactants, A and B for example, reactant B with a smaller and less attractive ΔH_f° may be preferable to A in view of greater quantities of products of larger negative values of ΔH_f°. Reaction performance, therefore, can be assessed directly only on the basis of heats of reaction; nevertheless, values of heats of formation provide some guide towards likely reactant types. Typical values of ΔH_r° are listed in Table 3.3 in descending order of propulsive attraction.

TABLE 3.3. Heat of Oxidation Reaction ΔH_r°

Fuel		M.W.	kcal/mole	CHU/lbm fuel
Hydrogen	$H_2(g)$	2	− 57·4	−28,700
Diborane	$B_2H_6(g)$	28	− 487·0	−17,400
Pentaborane	$B_5H_9(g)$	64	−1038·0	−16,200
Methane	$CH_4(g)$	16	− 210·8	−13,200
Ethane	$C_2H_6(g)$	30	− 368·4	−12,280
Acetylene	$C_2H_2(g)$	26	− 312·0	−12,000
Benzene	$C_6H_6(l)$	78	− 783·4	−10,040
Ammonia	$NH_3(g)$	17	− 97·1	− 5720
Methanol	$CH_3OH(l)$	32	− 170·9	− 5340
Hydrazine	$N_2H_4(l)$	32	− 151·0	− 4720

Representative values derived from ref. 2 and other sources.
1 CHU/lbm = 4186·8 J/kg.

3.1. Determination of Heats of Formation and Reaction

Where possible, thermochemical data are measured directly by experiment using calorimetric techniques. In other cases, indirect determination is effected by difference or by summation. In equation (2.3), for example, solution is possible if one quantity only is unknown, hence the heat of formation of a reactant may be calculated from the measured heats of reaction and the known heats of formation of the reaction products. The heat of formation of propane, for example, can, in fact, be measured experimentally and is found to be −24·8 kcal/mole. Since the heat of reaction of propane with oxygen is also measurable, together with the heats of formation of the oxidation products CO_2 and H_2O, the heat of formation of propane can also be determined arithmetically by difference, as follows:

$$C_3H_8(g) + 5O_2(g) \longrightarrow 3CO_2(g) + 4H_2O(l).$$
$$\Delta H_r^\circ = -530{\cdot}60 \text{ kcal/mole}$$

From equation (2.3),

$$\Delta H_r^\circ\ C_3H_8(g) = \sum \Delta H_f^\circ \text{ products } - \sum \Delta H_f^\circ \text{ reactants.}$$

Hence,
$$\begin{aligned} \Delta H_f^\circ\ C_3H_8(g) &= \sum \Delta H_f^\circ \text{ products } - \Delta H_r^\circ\ C_3H_8(g) \\ &= [3\,\Delta H_f^\circ\ CO_2(g) + 4\,\Delta H_f^\circ H_2O(l)] - \Delta H_r^\circ C_3H_8(g) \\ &= [3(-94{\cdot}1) + 4(-68{\cdot}3)] - (-530{\cdot}6) \\ &= -24{\cdot}9 \text{ kcal/mole.} \end{aligned}$$

An alternative approach is the summation of the contributions of the various atomic groups or radicals. The propane molecule in the above example may be represented structurally as:

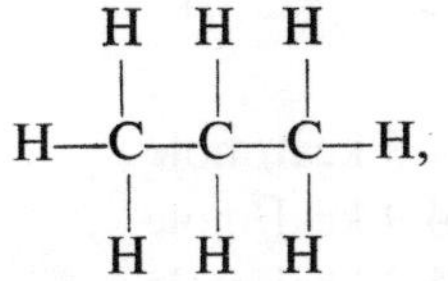

or CH_3—CH_2—CH_3.

Using representative values of radical contribution, the summation appears as:

$$\begin{aligned} \Delta H_f^\circ \text{ contribution for } CH_3 \text{ radical} &= -10{\cdot}1 \text{ kcal/mole} \\ \Delta H_f^\circ \text{ contribution for } CH_2 \text{ radical} &= -\ \ 5{\cdot}0 \text{ kcal/mole} \\ \Delta H_f^\circ \text{ contribution for } CH_3 \text{ radical} &= -10{\cdot}1 \text{ kcal/mole} \\ \text{Thus, } \Delta H_f^\circ \text{ for } C_3H_8(g) &= \overline{-25{\cdot}2} \text{ kcal/mole} \end{aligned}$$

A further method of determination utilizes the values of atomization heat and bond energy, typical values of which are listed in Table 3.4. Care must be taken to make corrections, where necessary, for resonance. The formation of propane may be written as follows:

$$3C(gr) + 4H_2(g) \xrightarrow[\Delta H_a \text{ added}]{} 3C(g) + 8H(g) \xrightarrow[D(X\text{—}Y) \text{ released}]{} C_3H_8(g)$$

Hence,
$$\begin{aligned} \Delta H_f^\circ\ C_3H_8(g) &= [3\Delta H_a C + 4\Delta H_a H_2] - [2D(C\text{—}C) + 8D(C\text{—}H)] \\ &= [3(135{\cdot}8) + 4(104{\cdot}2)] - [2(69{\cdot}3) \quad + 8(88{\cdot}5)] \\ &= 824{\cdot}2 - 846{\cdot}6 \\ &= -22{\cdot}4 \text{ kcal/mole.} \end{aligned}$$

TABLE 3.4. Heats of Atomization, and Bond Energies, kcal/mole

Process	ΔH_a
$C(gr) \longrightarrow C(g)$	135·8
$H_2(g) \longrightarrow 2H(g)$	104·2
$O_2(g) \longrightarrow 2O(g)$	118·3
$N_2(g) \longrightarrow 2N(g)$	225·6

Bond	$D(X—Y)$
C—C	69·3
C=C	106·4
C≡C	140·9
C—H	88·5
H—O	110·6
C—O	78·4
C=O	150 approx.

Representative values derived from various sources.

In summary, the approximate values of ΔH_f° $C_3H_8(g)$ obtained by the different methods appear as follows:

Experiment	−24·8 kcal/mole
Difference	−24·9 kcal/mole
Summation	−25·2 kcal/mole
Bond energy	−22·4 kcal/mole

but closer agreement would be expected with more accurate data taken from the latest literature.[3]

The bond energy concept also indicates clearly the marked increase in formation energy available should the atomization heat be eliminated and the reactions made to proceed from the elemental gaseous atoms. In the propane formation example above, a gross energy release of 846·6 kcal/mole would result instead of the net value of 22·4 kcal/mole, a forty-fold increase. This compares favourably also with the oxidation energy release of 530·6 kcal/mole. The storage of free atoms and other free radicals is extremely difficult in practice, however, without deep freezing or some elegant physical or chemical technique. Quantities of free atoms of oxygen and nitrogen are naturally available in the outer layers of the Earth's atmosphere, but in a widely dispersed state.

The exothermic decomposition figures for hydrogen peroxide given earlier may also be determined from the heats of formation. Using the difference method for the following decomposition:

$$H_2O_2(l) \longrightarrow H_2O(l)+\tfrac{1}{2}O_2(g)$$

gives

$$\begin{aligned}\Delta H_r^\circ\, H_2O_2(l) &= \Delta H_f^\circ H_2O(l)-\Delta H_f^\circ H_2O_2(l)\\ &= (-68\cdot 3)-(-44\cdot 8)\\ &= -23\cdot 5 \text{ kcal/mole.}\end{aligned}$$

As indicated in the previous chapters, the actual proportion of ΔH_r released depends upon the pressure and temperature of reaction, and the degree of dissociation at these conditions. The calculation of propellant performance during expansion through the nozzle was seen to be simplified by assuming that chemical equilibrium remains frozen at chamber conditions. Alternatively, gas composition may be assumed to shift to give chemical equilibrium at every point as both pressure and temperature fall progressively through the nozzle. Isentropic gas flow is assumed, and the local value of pressure calculated by iteration, allowing for the changes in molecular weight, to give the constant value of entropy required. Values of all other parameters related to the particular point of interest then follow. Computer techniques make such calculations practicable.

Frozen equilibrium gives a calculated performance some 3 to 10 per cent lower than that using shifting equilibrium, hence a slight factor of safety is incurred by using the former as a basis of performance prediction. The actual processes within the nozzle lie somewhere between these two limits.

3.2. High-energy Chemical Reactions

With chemical reactants, the prime requirement is a plentiful release of energy, and the discussion of heats of formation and bond energies showed that, for elemental fuels, values of oxidation energy per unit fuel-mass peak towards the lower limit of molecular weight (Fig. 2.9). Some of these elements exist normally in the solid phase, and others as gases. The influence of handling characteristics is discussed elsewhere,[12] but it is clear that service problems arise in such cases. Solids are difficult to feed to a combustion chamber, and can be stored only within it (see

Section 3.5). Gases are not feasible as such because of the extensive requirements for storage volume in the vehicle, and must be handled as cryogenic liquids with provision for venting the storage system and for minimizing vaporization loss by insulation, pre-cooling, etc. Liquids of low vapour pressure are classed as storable, and can be retained for

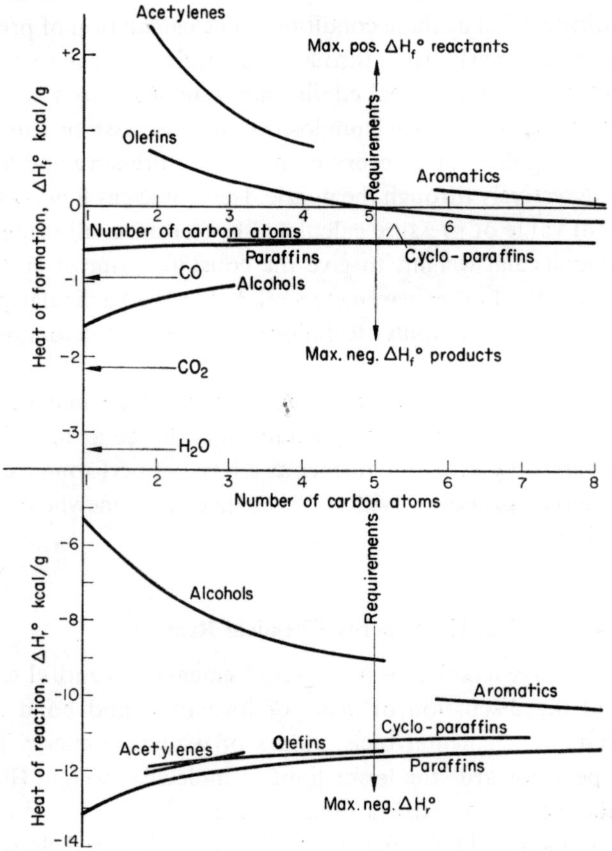

FIG. 3.1. Heats of formation, and reaction, of hydrocarbons. ΔH_f values derived from ref. 17. ΔH_r values derived from ref. 2.

extensive periods of time in vented systems. If vaporization due to chemical change is unlikely to occur with time, these liquids are classed as packageable and can be retained in sealed systems.

The combination of elements to produce compound reactants, therefore, gives a twofold advantage, since the molecular structure may be

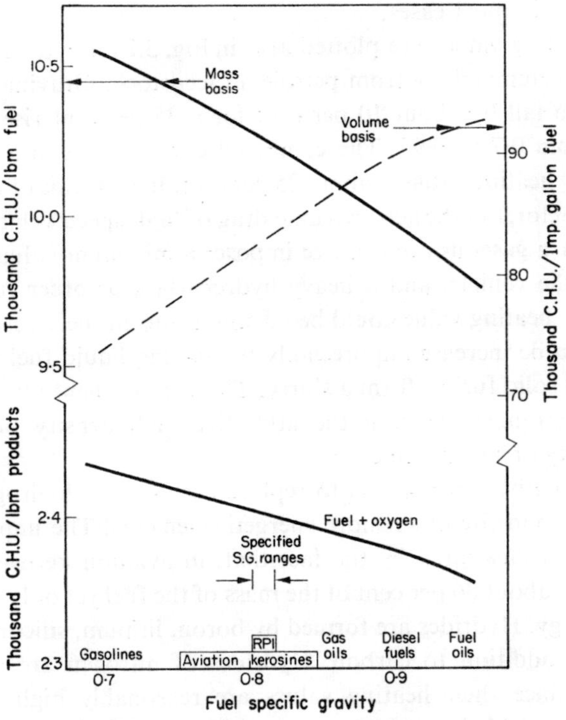

FIG. 3.2. Heating values of typical petroleum fuels.

selected both to liberate energy on decomposition, which is in addition to the oxidation energies of the component elements, and to give handling properties which are more convenient than those of the elements themselves. The hydrocarbons are a clear example of the latter, since hydrogen alone is not easily stored, and carbon alone is difficult to pump. The values of heats of formation of the first members of the most common hydrocarbon series are shown in the upper section of Fig. 3.1, together with corresponding values for the products H_2O, CO_2 and CO. The lower section of this figure shows the resulting values of heats of reaction (fuel heating values), taking into account the proportions of molecules involved. Again, better performance results from the lighter compounds in most cases.

Fuel heating values are plotted also in Fig. 3.2 for hydrocarbon fuels produced commercially from petroleum feedstock. Gravimetric values are seen to fall by about 10 per cent for a 35 per cent rise in specific gravity from 0·7 to 0·95. The combined effects result in a rise in the volumetric heating value of about 25 per cent. In all forms of flight, mass is a vital factor, but the aerodynamic drag of high-speed vehicles operating within a gaseous atmosphere imposes a minimum volume requirement on the vehicle, and a heavy hydrocarbon or other fuel of high volumetric heating value could be advantageous in these cases. Specific gravity can be increased appreciably by loading liquid fuel with finely pulverized solid fuel to form a slurry. This arrangement offers a practicable compromise between the attractive high density but difficult pumpability of solids alone.

An alternative approach is to replace carbon as a hydrogen carrier in hydrocarbons by other more energetic elements. The importance of this point is illustrated by the fact that, in aviation kerosine, carbon contributes about 86 per cent of the mass of the fuel yet only 62 per cent of the energy. Hydrides are formed by boron, lithium, silicon and magnesium in addition to carbon. Hydrides of nitrogen are also worth attention since their heating values are reasonably high, molecular weights reasonably low, and the nitrogen can be utilized as a temperature-controlling diluent.

Figure 3.3 illustrates dramatically the extension of the gravimetric and volumetric heating value relationships for both the hydrocarbon (i.e. carbon hydride) and boron hydride groups of fuels over the wider range of specific gravity. As indicated in Chapter 2, reaction energy based upon product mass is more appropriate for the rocket vehicle carrying all its own reactants, and these values have been included in Figs. 3.2 and 3.3. As expected from comparable values in Fig. 2.9, the boron hydrides are more attractive than the hydrocarbons, and the heavier molecular weights are favoured in this case.

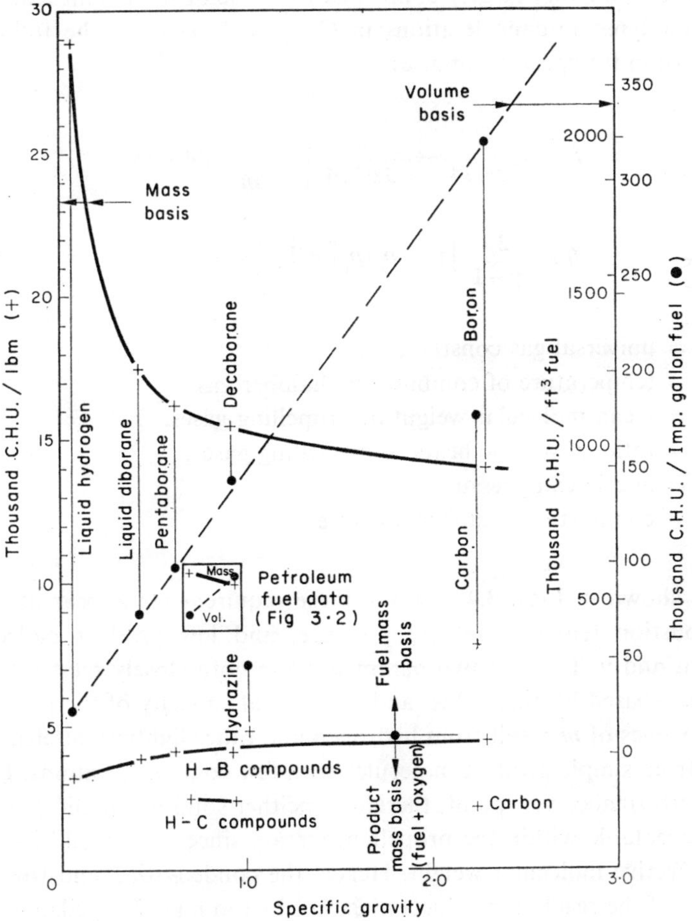

FIG. 3.3. Heating values of petroleum and boron hydride fuels.

3.3. Performance Requirements for Chemical Propellants

The foregoing discussion of bond energy changes via combustion, and of thermal-to-kinetic energy conversion via expansion, illustrates broadly the mechanism of chemical propulsion. Since the propellants themselves are generated as products of the combustion reaction, their properties are very largely controlled by those of the parent reactants. Thermodynamic considerations in Chapter 1 result in the following expression for specific impulse:

$$I_s = \frac{v_e}{32{\cdot}174} = \frac{1}{32{\cdot}174}\sqrt{\frac{GT_c\theta}{\boldsymbol{m}}}\ \text{lbf.s/lbm},$$

where $$\theta = \frac{2\gamma}{\gamma-1}\left[1-(P_e/P_c)^{\frac{\gamma-1}{\gamma}}\right].$$

G = universal gas constant,
T_c = temperature of combustion chamber gases,
$\boldsymbol{m}$ = mean molecular weight of propelling gases,
γ = ratio of specific heats of propelling gases,
P_e = nozzle exit pressure,
p_c = combustion chamber pressure.

As shown in Figs. 3.4 and 3.5, these requirements appear as high-combustion temperature and pressure, and low product molecular weight and γ. The first two parameters are both closely related to the product-based heating value, and to the heat capacity of the products. Low values of $\boldsymbol{m}$ result from hydrogen and other lightweight elements, also from simple product molecules with few component atoms. From the performance viewpoint, therefore, neither solid nor liquid particles are acceptable within the propellant stream since they would increase the effective molecular weight. Hence, the condensation and freezing-points of the reaction products must not be too high. Propellant-borne solids and liquids are also undesirable in practice due to nozzle fouling and erosion difficulties.

The value of γ, on the other hand, has little effect and does not present a practicable controlling variable. Nevertheless, the thermal conditions within a rocket nozzle are severe, as shown in Fig. 3.6, and the thermodynamic behaviour of the propellant gases in their passage at high temperature through the nozzle is significant. Hence, accurate assessment of chemical rocket performance must be based upon the resultant thrust

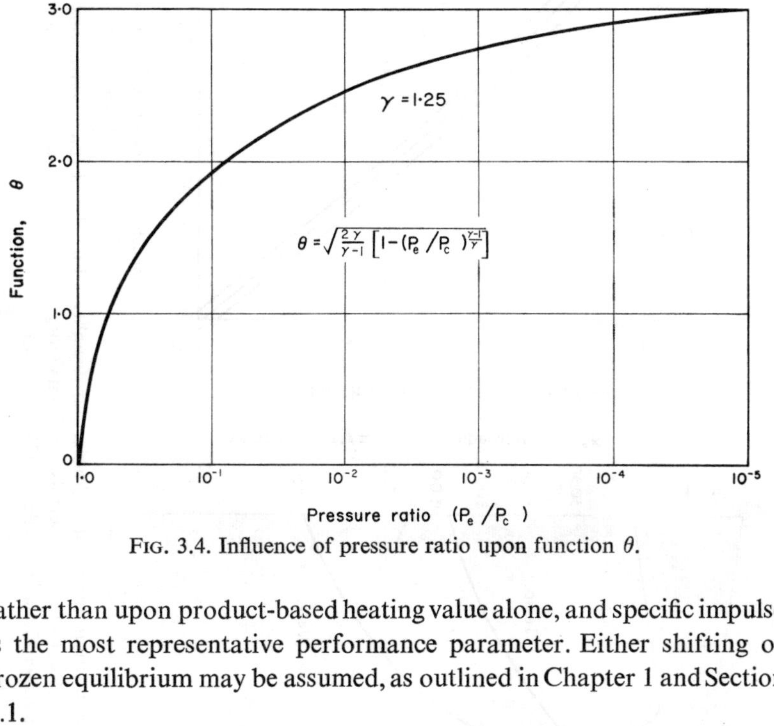

FIG. 3.4. Influence of pressure ratio upon function θ.

rather than upon product-based heating value alone, and specific impulse is the most representative performance parameter. Either shifting or frozen equilibrium may be assumed, as outlined in Chapter 1 and Section 3.1.

In practice, mixture ratio becomes a major operating variable, and Fig. 3.7 shows resulting variations in T_c and $\boldsymbol{m}$ together with the derived I_s, for two typical reactant pairs. Generally, although the maximum non-dissociated temperature occurs at the stoichiometric ratio, the influence of dissociation, together with the relative values of product heat capacities, is such that the maximum dissociated temperature lies on the fuel-rich side of stoichiometric. With fuels of high hydrogen con-

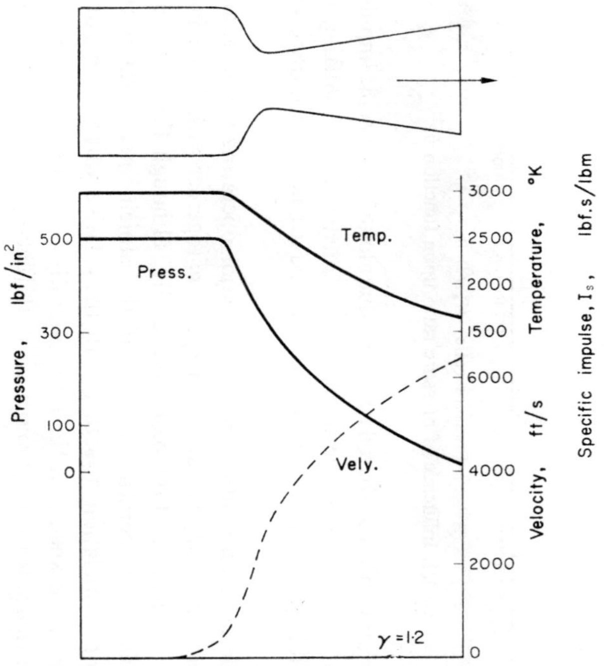

FIG. 3.5. Influence of (combustion temperature/molecular weight) ratio upon specific impulse. (Reproduced by permission from George P. Sutton's *Rocket Propulsion Elements*, Wiley, 1956—ref. 24.)

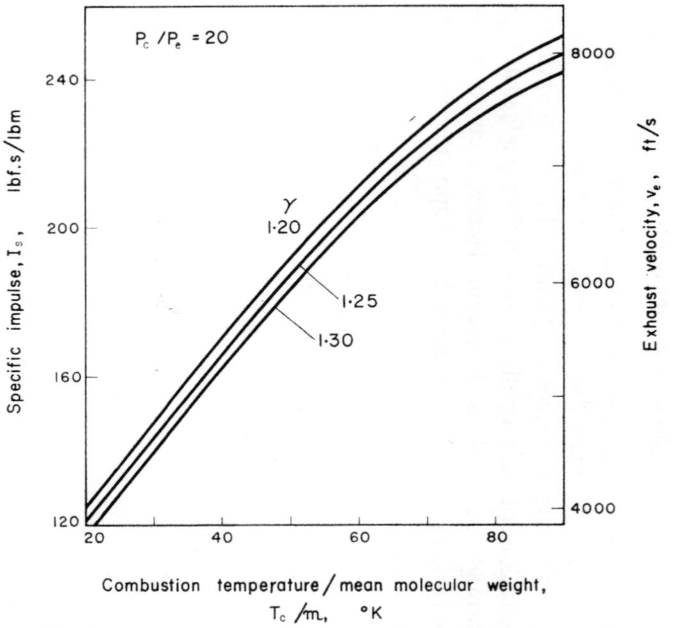

FIG. 3.6. Typical variations of temperature, pressure and velocity in a rocket nozzle.

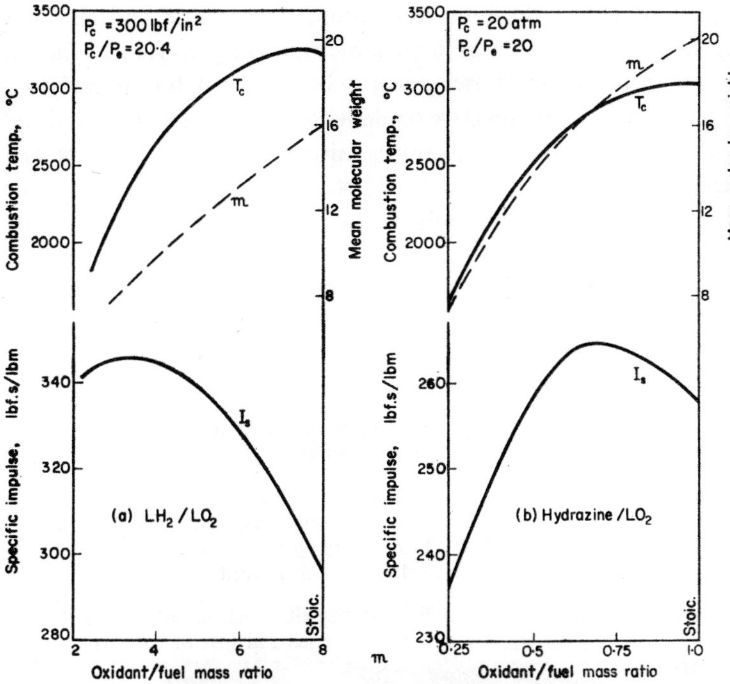

FIG. 3.7. Mixture ratio performance curves. (Reproduced by permission from George P. Sutton's *Rocket Propulsion Elements*, Wiley, 1956—ref. 24.)

tent, therefore, producing large concentrations of water vapour, the reducing value of ***m*** with fuel enrichment brings the specific impulse peak further over towards the fuel-rich side. In the following comparison of reactant performance, peak values of specific impulse are used as indicative of the optimum performance available from each reactant pair.

A knowledge of the high-energy-producing reactions available, together with the properties desired in the resulting products, leads to the selection of those reactants which produce both the required intensity of energy and the necessary expansion behaviour in the nozzle.

The overall effects of low molecular weight suggest hydrogen as one of the principal fuels for rocket engines. It can be liquefied for storage, but the specific gravity at its atmospheric boiling point of −253° C is only 0·071, so that tankage volume needs to be high. Of the other fuel elements, beryllium, lithium, aluminium, magnesium and silicon are likely to produce oxides of high molecular weight. Hence, the most likely elements for rocket use appear to be hydrogen, boron, carbon and nitrogen, together with reactive oxidants such as oxygen itself and fluorine. The various compound forms also available provide a broad grouping of six types of chemical reactant, as shown in Table 3.5. Since volume limitations do not apply in space, no further discussion is given

TABLE 3.5. Grouping of Chemical Reactants

Type	Typical members	Main features
Free radical	H, OH, NH, CH	Very high energy release Highly unstable
Elemental	H, Be, B, C	High energy release Gases require excessive storage volume Liquids cryogenic Solids difficult to feed
Hydrocarbon (i.e. carbon hydride)	Kerosine ($C_{12}H_{24}$ approx.)	High availability, reasonable performance and ease of handling
	Methane, CH_4	Highest energy release of hydrocarbons, but gaseous
Non-carbon hydride	B_2H_6, B_5H_9, $B_{10}H_{14}$ (also hydrides of Be and Li)	High energy release with reasonable ease of handling
	NH_3, N_2H_4	Reasonable energy release and low molecular weight
Slurry	B, Mg or Al suspended in kerosine	High energy release for volume-limited applications in Earth's atmosphere
Solid	Nitroglycerine + nitrocellulose	Reactant pairs mixed and/or compounded into single medium Storage effected within combustion chamber "One-shot" operation

here on the slurries. The two main types of practical reactant, liquid and solid, are discussed in the following sections.

3.4. Liquid Reactants

Of the remaining reactant types shown in Table 3.5, only the hydrides provide fuels in the liquid phase at normal temperatures and pressures, whereas hydrogen can be handled as a cryogenic liquid with consequent penalties of handling cost and insulation mass. Since liquid oxygen is also a cryogenic material, oxygen-rich compounds can prove useful alternatives. These include concentrated hydrogen peroxide (high test peroxide, or HTP), di-nitrogen tetroxide (N_2O_4), and concentrated

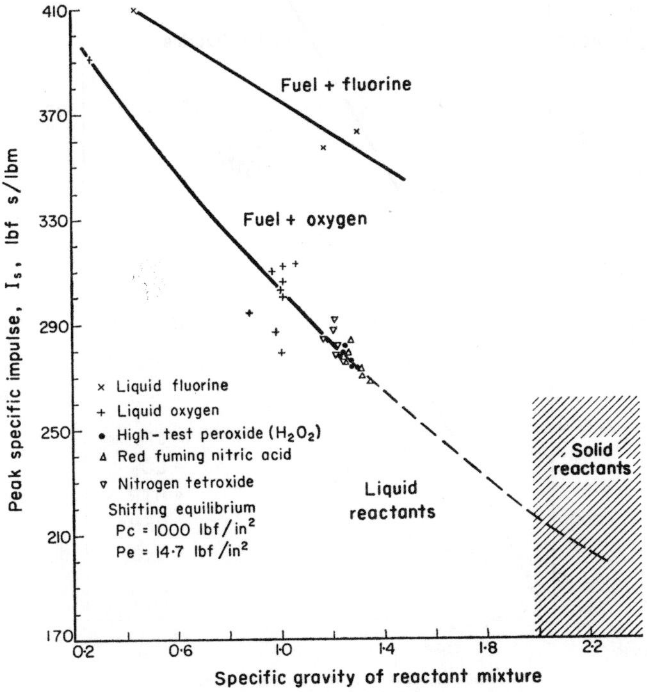

FIG. 3.8. Relationship between specific impulse and specific gravity (values from ref. 4).

nitric acid with up to either 1 per cent or 25 per cent of di-nitrogen tetroxide (white fuming or red fuming respectively).

Although specific impulse is the more accurate representation of propellant performance, a general direct relationship is to be expected between specific impulse and product-based heating value, and Fig. 3.8 indicates a similar trend of high values of peak specific impulse with low specific gravity of reactant mixture. Because of the inclusion of

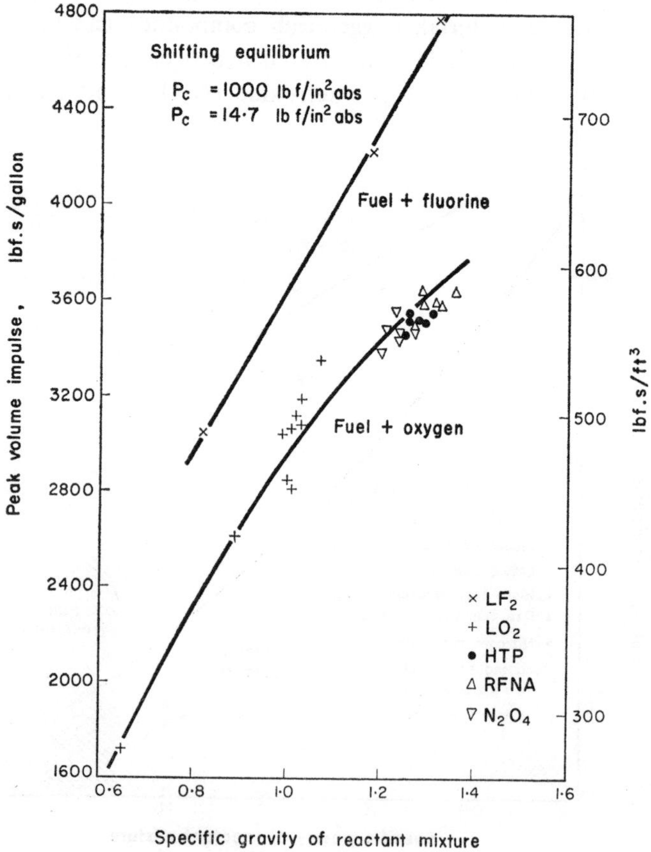

FIG. 3.9A. Relationship between peak volume impulse and specific gravity (small-scale plot) (values from ref. 4).

specific gravity in the impulse term, a plot of volume impulse shows a closer direct relationship with specific gravity, as in Fig. 3.9A. The large-scale replot in Fig. 3.9B indicates a difference of about 8 per cent between values of shifting and frozen equilibrium.

The relative performance of reactant pairs can be illustrated broadly

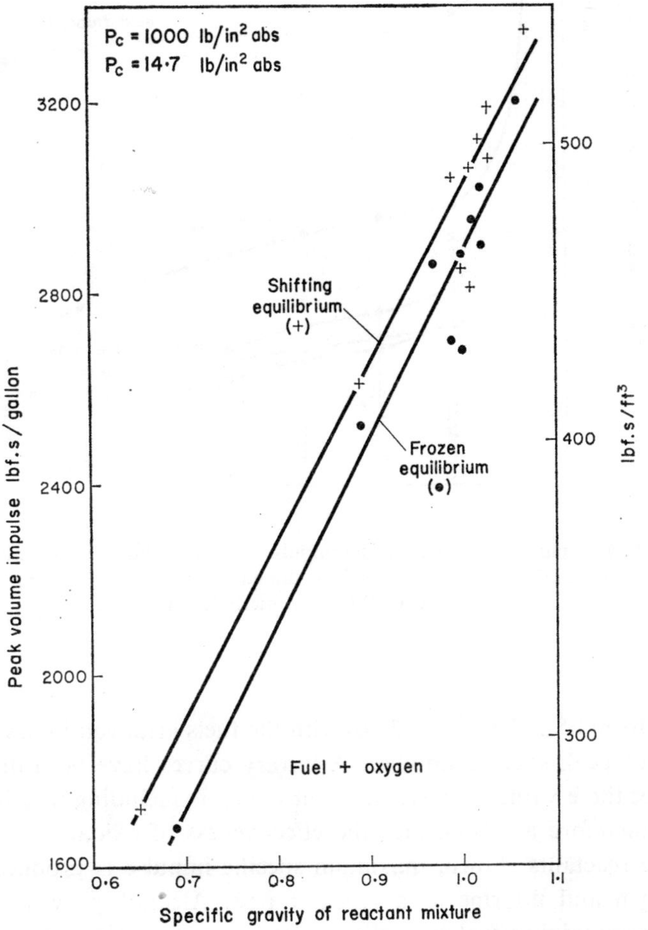

FIG. 3.9B. Relationship between peak volume impulse and specific gravity (large-scale plot) (values from ref. 4).

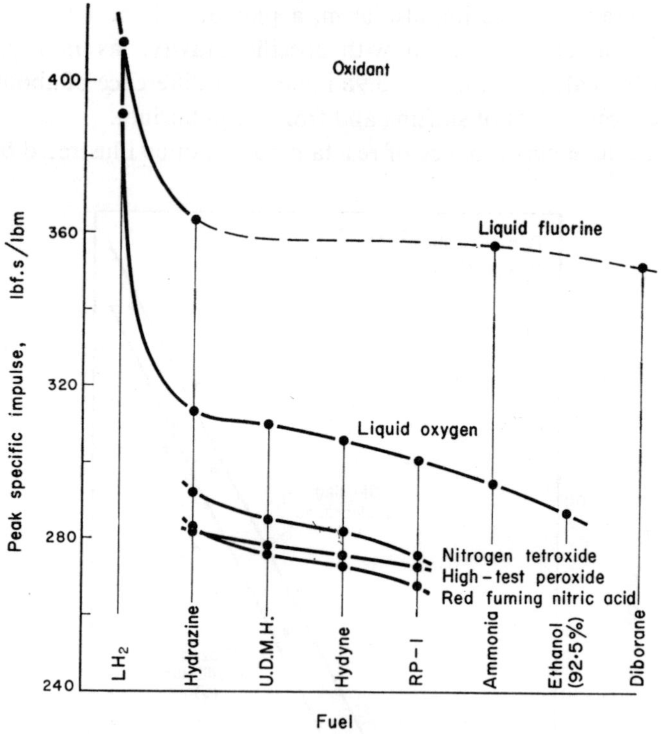

FIG. 3.10. Comparison of peak specific impulse values for liquid reactants (values from ref. 4). UDMH = unsymmetrical dimethyl hydrazine. HYDYNE = 60/40 mass mixture UDMH and diethylenetriamine.

in the form of a chart (Fig. 3.10) with the fuels arranged in descending order of peak specific impulse. Arbitrary curves have been drawn to connect these values for five different oxidants including fluorine. The chart therefore also indicates the effectiveness of oxidants, and that, for the reactants shown, maximum specific impulse is produced with hydrogen and fluorine as a reactant pair. Alcohol proved a useful oxygen-containing fuel in earlier years when operating temperatures were more strictly limited, but improved materials and design tech-

TABLE 3.6. Representative Properties of Liquid Reactants[12]

Property \ Material	Hydrocarbons				LH_2	60/40 Boron AVTUR slurry	Hydrides					Oxidants			
	AVTUR	RP-1	Poly-cyclic naph-thenes	Ethanol			B_2H_6	B_5H_9	NH_3	N_2H_4	UDMH	LF_2	LO_2	HTP	HNO_3
Lower heating value CHU/lbg	10,250	10,280	10,270	6650	28,700	12,500	17,400	16,200	5720	4720	7870				
Lower heating value CHU/gal	82,000	83,000	88,900	52,700	20,090	164,000	74,800	98,800	35,000	47,700	61,400				
Sp. gr. at 60/60 °F	0·8	0·8	0·866	0·792	0·071 b.p.	1·31	0·43 b.p.	0·61	0·61	1·01	0·78	1·51	1·14 b.p.	1·35	1·56
Sp. heat at 60° F CHU/lb° C	0·50			0·61	2·33 b.p.		0·48	0·57	1·12	0·75	0·65	0·37	0·4 b.p.	0·61	0·42
Vaporization heat CHU/lb	67			204	109		125	122	326		134	41	51	346	115
Boiling point ° C	150/300	185/274	201/275	80	−253	150+	−93	63	−33	114	63	−188	−183	152	86
Freezing point ° C	−40	−40	−60	−118	−259	−40	−165	−47	−78	2	−58	−220	−220	−22	−42
Flash point ° C	37·8	43·3		21							1				
Spontaneous Ign temp ° C, in air at 1 atm	254			439	570				651	52	250				
			Typical												

niques permit the use of more energetic reactants. The general properties of liquid-chemical reactants are shown in Table 3.6.

3.4.1. Combustion Characteristics

In continuous-flow air-breathing combustion systems, the stabilization of flame is generally dependent upon rapid vaporization of the liquid-fuel spray, and the rapidity of the mechanical mixing and chemical reaction of the fuel vapour and air. Desirable fuel properties, therefore, include low viscosity, surface tension and heat capacity, with high volatility, thermal conductivity and flame speed, and the chamber design incorporates a baffle system to ensure vigorous mixing and to provide a sheltered zone for anchoring the flame. With rocket engines, however, both reactants are injected into the chamber, and stability is generally achieved through the flow pattern resulting from spray impingement. Commonly, fuel is used as a nozzle coolant, the heat extracted from the nozzle walls thus being circulated back to the chamber. In some cases, this heat, together with the high level of injection pressure, is sufficient to raise the condition of the reactant above the critical point such that no latent heat is required, and vaporization takes place virtually simultaneously on injection into the chamber.

High rates of chemical reactivity are also desirable in order to ensure both rapid ignition of the impinging reactants, and rapid subsequent combustion. The former is particularly important at start-up since any delay will give rise to a "hard" start of the reactant-filled chamber, and may lead to explosion and failure. The latter permits the length, and therefore weight, of the combustion chamber to be minimized. Under laminar atmospheric conditions, flame speeds of the hydrocarbons do not exceed about 4 ft/s, but are increased considerably at high pressures, under turbulent conditions, and in oxygen. The relatively high flame speeds of hydrogen and the boron hydrides, therefore, are attractive.

Fuels burning with high luminosity, caused by glowing particles in the flame, are liable to form smoke and to promote metal fatigue due to radiant heat transfer to the combustor walls. The aromatics fall into this category. One valuable property which helps ensure ignition, flame stability and relight at altitude is the spontaneous ignitibility of the reactants on contact in the liquid phase. This phenomenon is termed "hypergolic" ignition, and is exhibited by hydrazine and HTP, aniline and nitric acid, and some other reactant pairs. It simplifies the detail design of the rocket engine by the elimination of electrical ignition equipment, but inevitably increases the risk of fire in handling. A valuable development is the system of "thermal" ignition, in which HTP is decomposed catalytically in the combustion chamber, the decomposition temperature being sufficient to ignite kerosine spontaneously. This preserves the convenience of spontaneous ignition within the chamber, without the increased fire danger outside it.

3.5. Solid Reactants

The concentration of propellant energy into a solid medium is attractive for economy of storage volume, but the medium must contain both members of the reactant pair, i.e. both fuel and oxidant, which must not tend towards spontaneous reaction. Hence the activation energy (the energy input necessary for the reaction to start) must be sufficiently high for storage stability. Solid fuels generally comprise such organic materials as hydrocarbons, polymers, plastics and rubbers, and solid oxidants include perchlorates, crystalline nitrates and other inorganics together with colloidal organic nitrates. In many cases, each component contains both oxygen and fuel elements.

Since solid materials do not lend themselves to flow transfer, storage is effected within the combustion chamber, and the charge "grain" is shaped such that combustion can progress at a controlled rate to completion. Typical charge sections are shown in Table 3.7, together with the corresponding variation of burning rate, i.e. burning area, with time. At a given pressure, the burning rate remains sensibly constant and, since the most common requirement is a constant thrust, the sectional geometry is selected to present a constant burning area as successive layers of material are consumed uniformly.

Burning rate is also affected by propellant composition, and can be controlled further by the use of additives. The storage requirement of

TABLE 3.7. Solid Charge Sections

DESCRIPTION	SECTION	TYPE OF BURNING	THRUST
Solid cylinder	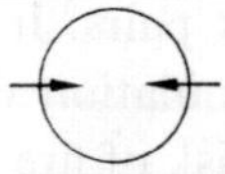	Regressive radial	Regressive
Inhibited tubular	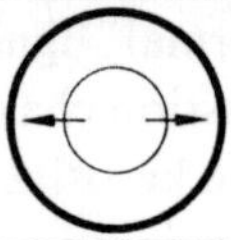	Progressive radial	Progressive
Tubular	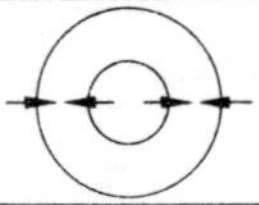	Combined radial	Approximately constant
Clover leaf		Outwards radial	
Star centre			
Cruciform		Inwards radial	
Cogged	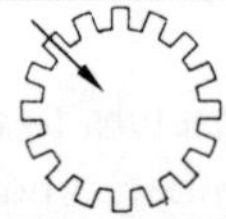		
Axial		Cigarette	

retaining shape and consistency over extended time periods and throughout changes in temperature and humidity impose stringent property requirements of a physical as well as a chemical nature. In the absence of storage tanks, pumps and control valves, solid reactant systems are intrinsically simpler than liquid reactant systems, but they do not offer flexibility of operation, and cannot easily be shut down and restarted. Furthermore, their thrust performance is lower than that of liquid reactants. Although the specific impulse parameter can be employed as a measure of propulsion performance, the "one-shot" operation makes more appropriate the total impulse parameter derived in Section 1.3.2.

3.5.1. Reactant Ballistics

The equilibrium pressure attained within the combustion chamber of a solid reactant rocket is the resultant of two independent relationships between pressure and burning rate. Treated as a pressure vessel exhausting gas through an orifice, mass flow rate from a rocket-nozzle system appears as follows:

$$\dot{m} \propto P_c A_T,$$

where P_c = chamber pressure and A_T = area of nozzle throat. Since the gases are provided by the uniform consumption of a charge, $\dot{m}$ = volume rate of burning×charge density,

$$= (ar)D,$$

where a = burning area, r = linear burning rate, and D = charge density. Thus the first relationship evolves as:

$$r \propto \left(\frac{A_T}{aD}\right) P_c,$$

i.e.

$$r \propto P_c.$$

The second relationship, observed from experiments conducted in a variety of combustion rigs over a wide range of pressure, appears as follows:

$$r \propto P_c^n,$$

and, for solid reactants, n is commonly equal to 0·5 but may lie between 0·4 and 0·85. This may be due to the presence of carbonaceous combustion products which act catalytically at low pressures but tend to be consumed at high pressures.

These two relationships combine as in Fig. 3.11 to give a condition of equilibrium operation at the intersection point. Any tendency towards higher pressure, for example, would produce a rise in the burning rate shown by the reactant curve, but the rocket curve indicates an increased rate of efflux corresponding to a relatively higher burning

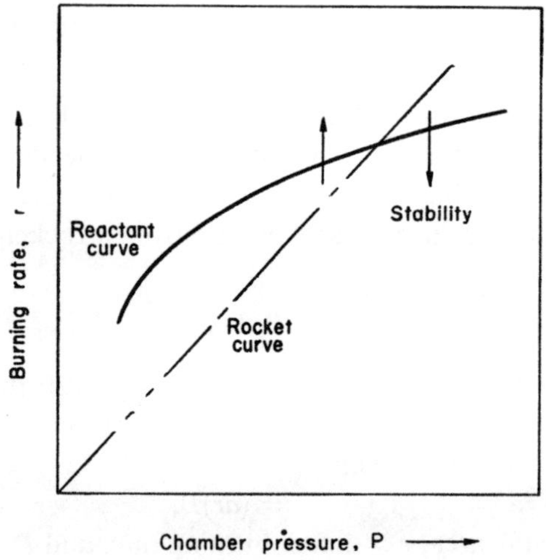

FIG. 3.11. Curves of burning rate vs. pressure.

rate, hence the system returns to its initial equilibrium. Stability of operation becomes a direct function of the difference between the gradients of the two curves about the equilibrium point. Since the rocket curve is fundamental to the system, stability requires the reactant curve to show the minimum pressure dependence. The incorporation of a lead salt additive in the reactant accentuates the curvature of the r–P relationship over a limited range, probably through increased catalytic action, leading to a "plateau", thus improving the stability. Reactants so treated are described as platonized.

Since the burning rate increases with temperature, the level of the reactant curve is raised. The rocket curve, however, will depress to a slightly lower gradient since higher pressures would result from higher temperatures. Consequently, for a given nozzle-throat area, the equilibrium point would vary with ambient temperature over a range of pressure as shown in Fig. 3.12, and would give a corresponding range of thrust. This may well be unacceptable for a precisely-controlled flight programme, hence the nozzle size may need to be selected to suit the

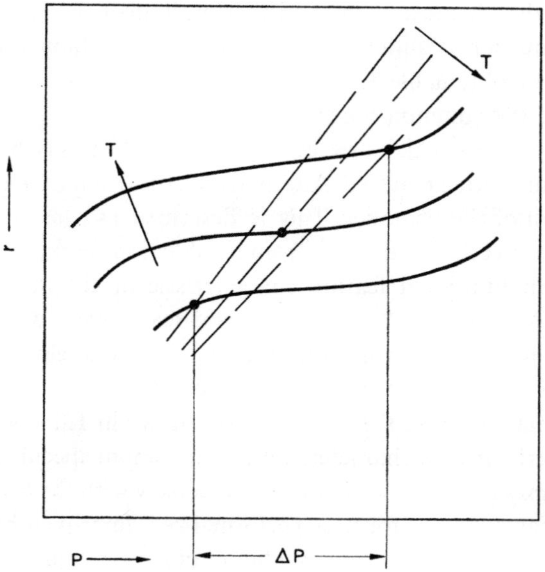

FIG. 3.12. Effect of temperature on burning rate–pressure curves.

ambient temperature. At extremes of temperature, the operating pressures may peak towards chamber bursting, or sink towards intermittent burning. The sudden drop in pressure caused by the piercing of the chamber wall, for example, is likely to quench the combustion completely. This effect can be used experimentally to measure the burning rates of a number of reactant samples simultaneously in one test. Constancy of reactant quality can be checked during production by measuring the burning time of a $\frac{1}{8}$ in. diameter strand of reactant burning axially over a 5 in. gauge length. In practice, however, the gas velocity past the burning surface of the charge grain causes an "erosive burning" effect due to the increased rate of heat transfer, and a more realistic assessment of burning rate under operating conditions is made by means of a small-scale grain in a rocket chamber-nozzle configuration. By incorporating a fusible nozzle which expels rapidly after a predetermined burning time, combustion ceases abruptly due to the sudden drop in pressure, as discussed above. Hence, a number of different charge grains can be loaded into the chamber to be ignited simultaneously by the ignition charge, and quenched simultaneously by the pressure blow-out. Burning rates are then obtainable from the dimensions of the unburnt grain portions and the common burning time.

Reactant surfaces which are required to burn last may be protected by layers of inhibiting material. Alternatively, the reactant may be bonded to the wall of the chamber. This is described as case bonding, and ensures the protection of the wall from combustion temperature during the firing. An additional feature characteristic of the design of solid reactant rockets is the "obturator" semi-seal device which permits pressure equalization throughout the chamber but eliminates flame contact with those surfaces selected for protection.

The four main types of solid reactant are shown in Table 3.8, together with their performance characteristics. A maximum specific impulse of 260 lbf. s/lbm is noted, which compares adversely with the 410 lbf. s/lbm for liquid reactants. Combustion performance is improved by the presence of other additives such as aluminium, beryllium and lithium. In particular, temperature control can be effected by the incorporation of ammonium picrate as a coolant. The widely-used cordite comprises 41 per cent nitrocellulose, 50 per cent nitroglycerine and 9 per cent carbamite as a binder, together with a small proportion of wax in order to lubricate the extrusion die. The black powder commonly used as an igniting agent comprises 61·6 per cent potassium nitrate, 23·0 per cent charcoal and 15·4 per cent sulphur, with clay, starch or glue as a binder.

Solid reactants are described as deflagrating (as distinct from detonating) explosives since they contain all the elements necessary for reaction but react by combustion alone rather than by a mutual combustion-and-shock wave, i.e. a detonation wave. Restriction of the outlet orifice

TABLE 3.8. Solid Rocket Reactants

Type	Typical examples	General properties	I_s lbf.s/lbm (1000 to 14·7 lbf/in^2)	Burning rate r, in./sec	$T°$ K
Single base colloidal	Nitrocellulose	Cool burning, hygroscopic Capable of detonation	–	–	–
Double base colloidal	Nitrocellulose plasticized with nitroglycerine Extruded or cast	Nitric esters capable of detonation, but stabilize each other Not case bondable	170–200	0·2–0·8	1900–3000
Composite	Pressed—black powder	Not case bondable	170–220	0·05–0·2	1600–2300
	Plastic—ammonium perchlorate with polyisobutene binder Polymerizable binder—rubbers	Case bondable	170–260	0·05–1·5	1400–3800
Composite modified double base	Inorganic oxidant with nitrocellulose binder	High performance Case bondable	240–260	0·3–1·0	> 4000

Notes: 1. Density varies from about 0·054 to 0·067 lb/in^3. (Specific gravity from about 1·95 to 2·42.)

2. Addition of aluminium, beryllium or lithium improves performance.

3. Commercial dynamite = nitroglycerine absorbed in sawdust.

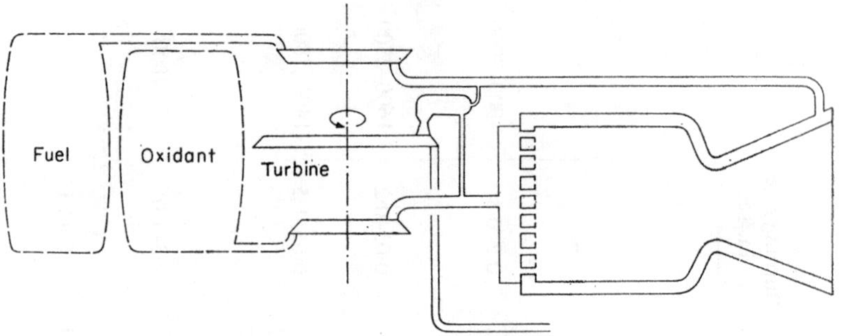

(a) Liquid bi-reactant

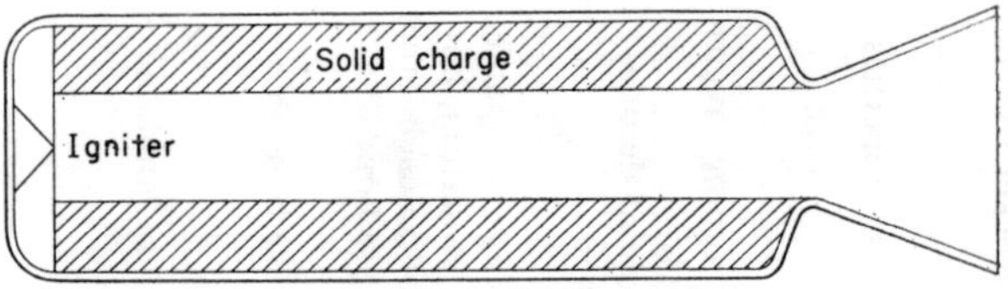

(b) Solid mono-reactant

FIG. 3.13. Schematic diagram of chemical rockets.

of a reaction chamber raises the equilibrium pressure, but the virtual closure of the orifice, as with a projectile in a gun, leads only to more rapid combustion and a pressure sufficiently high to propel the projectile without giving rise to detonation and resultant shattering of the gun barrel.

Line diagrams of typical chemical rockets for use with liquid and solid reactants are shown in Fig. 3.13.

Symbols

A_T = Area of nozzle throat
a = Burning area of solid reactant
D = Density of solid reactant

$D(X—Y)$	=	Bond energy to dissociate X and Y at datum conditions
G	=	Universal gas constant
ΔH_a°	=	Heat of atomization at datum conditions
$\Delta H_{\text{cond}}^\circ$	=	Heat of condensation at datum conditions
ΔH_f°	=	Heat of formation at datum conditions
ΔH_r°	=	Heat of reaction at datum conditions
$\dot{m}$	=	Mass flow rate of propellant
m	=	Mean molecular weight of propellant
P	=	Pressure
r	=	Linear burning rate of solid reactant
T_c	=	Static temperature of propellant at reaction chamber exit plane
v_e	=	Effective velocity of propellant through nozzle exit plane
γ	=	Ratio of specific heats of propellant
θ	=	Function of pressure and γ.

Subscripts

c	=	Reaction chamber
e	=	Nozzle exit

CHAPTER 4

NUCLEAR PROPULSION

NUCLEAR reactions are seen to offer very substantial quantities of energy. In terms of unit mass of material consumed, the complete oxidation of one pound of stoichiometric hydrogen-oxygen mixture, for example, releases 3188 CHU, whereas the complete (if this were possible) conversion of one pound of material into energy represents 20×10^{12} CHU, giving a factor of improvement of over 6×10^{9}. Currently, no more than about 2 per cent of the energy equivalent of mass is obtainable as "nuclear energy". Some of the more important factors influencing the release of nuclear energy are discussed below, followed by a review of the methods of application to rocket propulsion.

4.1. Radioactivity

The nuclei of elements up to atomic number 82 (lead) and mass number 209 are stable, but heavier nuclei are radioactive and tend to decay spontaneously by the emission of particles. The probability of an unstable nucleus decaying within a given time is constant and, for a unit period of time, is known as the decay constant, λ. For a number (N) of like radioactive nuclei in a given mass, the rate of disintegration is known as the activity (I) and is measured in curies, where 1 curie = $= 3{\cdot}7 \times 10^{10}$ disintegrations per second. Hence:

Activity $= I =$ rate of decay of radioactive nuclei $= -dN/dt = \lambda N$ disintegrations per second.

Mathematical integration gives the variation of N with t:

$$\int_{N_o}^{N} dN/N = -\int_{o}^{t} \lambda\, dt = \ln N/N_o = -\lambda t,$$

and $$N = N_o e^{-\lambda t}\,. \tag{4.1}$$

This exponential decay is shown in Fig. 4.1.

Since radioactivity continues exponentially, the theoretical total life of all radioactive nuclei is infinite. However, the shape of the curve for each type of elemental nucleus is a function of λ, and is different for each element. The concept of the half-life, $t_{\frac{1}{2}}$, as the time taken to reduce

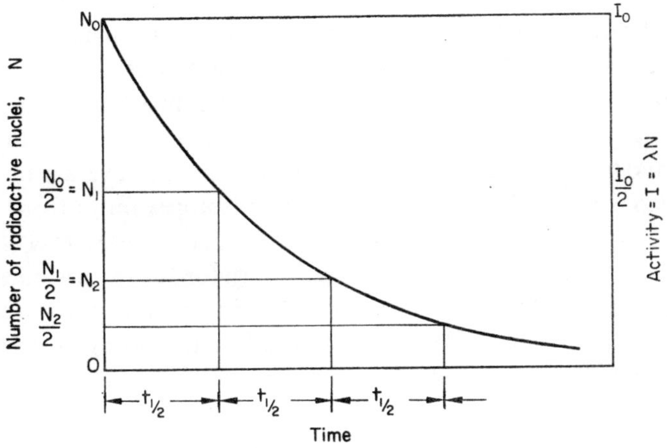

FIG. 4.1. Exponential decay and half-life.

the number of radioactive nuclei (and also the activity) by half, therefore, provides a means of identifying the elemental nature of the nuclei. The half-life is constant irrespective of the initial number of nuclei in a given mass, i.e. irrespective of the instant in time when measurement starts. Being a nuclear characteristic, the half-life is independent of any chemical compounding of the element, or of conditions of temperature and pressure. Since $t = \dfrac{\ln N_o/N}{\lambda}$ from equation (4.1)

$$t_{\frac{1}{2}} = \frac{\ln 2}{\lambda} = \frac{0{\cdot}693}{\lambda}\ \text{s}.$$

Half-lives for radioactive isotopes range from about $1{\cdot}39 \times 10^{10}$ years for Th_{90}^{232} down to about 3×10^{-7} seconds for Po_{84}^{212}. Half-lives of the two main isotopes of uranium, U_{92}^{235} and U_{92}^{238}, are $4{\cdot}5\times10^{9}$ years and $7{\cdot}1\times10^{8}$ years respectively.

Most elements of mass number above 220 are fissionable by particle bombardment. The additional, incident, energy induces vibrations which tend to disturb the spherical geometry of the nucleus. The electrostatic migration of the charged protons to the more sharply curved surfaces of the disturbed nucleus accentuates the disturbance and assists the tendency to partition. The nucleus accordingly divides into two fission fragments which, for uranium, are most likely to have mass numbers of about 95 and 140. Also released are two or three neutrons which may contain energy appropriate for fission processes with other nuclei and hence lead to sustained chain reactions. The average numbers of neutrons released at fission for uranium isotopes range from about 2·46 to 2·54. In order to sustain a chain reaction, the energy level of these neutrons must lie within the ranges necessary for further fission which, for the uranium isotopes, are illustrated in the following sections. Neutrons are also released as a result of radioactive decay, as described above, and these "delayed" neutrons are a valuable aid in reaction control.

4.2. Inelastic Collisions

The bombardment of matter with atomic structural particles gives rise to collisions between the incident particles and the target nuclei. Neutrons are commonly used as incident particles in order to provide reasonable mass without the decelerating effect of electrostatic repulsion. The nature of these collisions is dependent upon the levels of energy involved, and is described as elastic if the particles and nuclei are merely deflected, the overall kinetic energy remaining constant following the laws of classical mechanics. With an inelastic collision, the particle is captured by the nucleus and hence changes the energy balance of the nucleus since the nuclear binding energy applies also to the additional particle, and since some kinetic energy may be associated with the incident particle itself. The resultant excitation of the nucleus may lead to ejection of either a particle or an energy wave, or to fission into different elemental nuclei of roughly comparable size. If the number of protons is changed, the elemental nature of the target matter is transmuted. The following shorthand notation is used for inelastic collisions:

$$X(a,b)\,Y,$$

where X is the target nucleus, a is the incident particle (commonly a neutron), b is the ejected particle(s), and Y is the resultant nucleus (or nuclei).

Transmutation by neutron bombardment, therefore, may be represented as

$$X_Z^A(n,p)\,Y_{Z-1}^A,$$

when a proton is ejected, or as

$$X_Z^A(n,\beta)\,Y_{Z+1}^{A+1},$$

when a neutron converts to a proton and an electron, the latter ejecting as β^- radiation.

Fission by neutron bombardment may be represented as

$$X_Z^A(n,xn)\,Y_{Z1}^{A1}+Z_{Z2}^{A2},$$

where x is the number of neutrons released, $A = A1+A2-(x-1)$, and $Z = Z1+Z2$.

The captured neutron, as shown earlier, assumes the binding energy of the target nucleons and also gives up its own kinetic energy. The binding energy was shown in equation (2.8), and can be expressed entirely in terms of numbers of nucleic particles in the following semi-empirical expression derived from the liquid drop concept:

$$B \text{ per nucleus} = 14{\cdot}01A - 12{\cdot}7A^{2/3} - \frac{0{\cdot}584Z^2}{A^{1/3}} - \frac{19{\cdot}3(A-2Z)^2}{A} - \frac{33{\cdot}5\delta}{A^{3/4}}, \quad \text{MeV}$$

and similarly $\overline{B} = B/A$.

The first term represents the energy required to overcome the nucleic attraction forces between the A nucleons, and the remaining terms reduce this respectively due to the nucleus surface-tension effect, the electrostatic repulsion of the protons, the loss of stability due to the excess of neutrons over protons, and the lower stability with odd values of protons (Z) and/or neutrons (N). Hence, $\delta = +1$ for Z and N odd, $= 0$ for either Z or N odd, and $= -1$ for Z and N even.

This expression applies satisfactorily to the heavier nuclei, so that the effect of an incident neutron upon a uranium nucleus can be calculated. The conversion of U^{235} to U^{236} by neutron capture, for example, results in the release of 6·7 MeV, the value of $\overline{B}$ for these nuclei.

The activation energy for fission may also be derived from the liquid drop model and can be expressed as

$$E_A = 0{\cdot}89A^{2/3} - 0{\cdot}02Z^2/A^{1/3}, \quad \text{MeV}.$$

Since this gives an E_A value of 6·6 MeV for U^{236}, it follows that the binding energy released by the neutron capture of U^{235} is more than sufficient to effect the fission of the resulting U^{236}. Hence, no initial kinetic energy of the incident neutron is necessary, and a "thermal" neutron (see below) is adequate.

In a similar manner, the binding energy released from neutron capture by a U^{238} nucleus is found to be 5·5 MeV, which is less than the 7·0 MeV activation fission energy of the resulting U^{239} nucleus. Hence, the balance of 1·5 MeV must be provided by the neutron in the form of kinetic energy; consequently U^{238} is fissionable only by fast neutrons.

With low-incident energy, consecutive elastic collisions lead to thermal equilibrium between the incident particles and the target nuclei. The use of low-energy neutrons as incident particles, therefore, leads to an equilibrium condition with "thermal" neutrons having energies of about 0·02 eV, corresponding approximately to a temperature of 20° C and a neutron velocity of 2200 m/s. A fission reaction, however, results in the emission of more energetic neutrons which may reach the fast range of 1 to 2 MeV.

Since many inelastic capture processes are possible, each reaction is associated with a probability of occurrence which is expressed in terms of the ratio of the effective area of the target nucleus to the total area of bombardment. The effective target area of a nucleus is known as the nuclear cross-section, and is measured in units of either cm^2 or barns, where 1 barn $= 10^{-28}$ m^2. The nuclear cross-section is a function

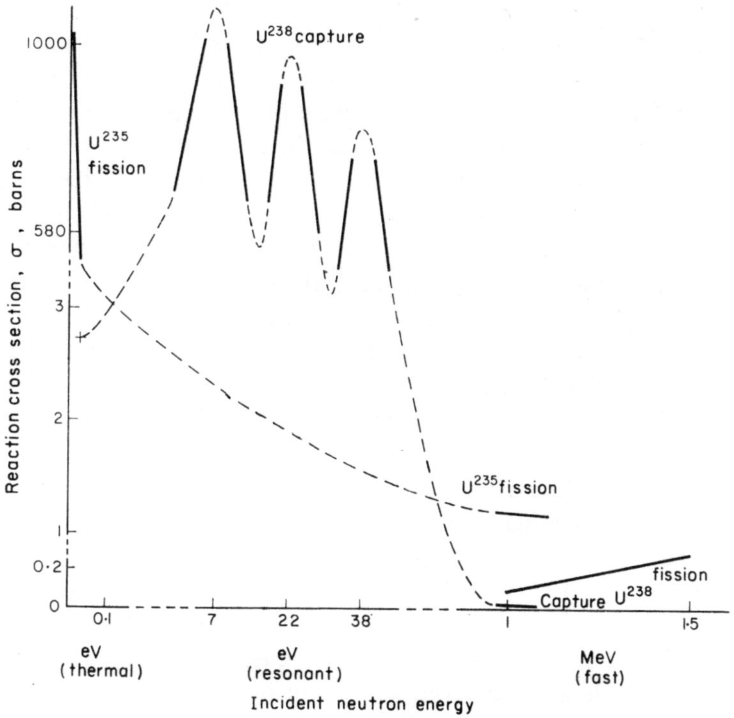

FIG. 4.2. Variation of fission and capture probability with incident neutron energy, for uranium 235 and 238 (derived from refs. 8, 9, 16 and 23).

of the type of reaction, which is dependent upon the incident kinetic energy. The nuclear cross-sections for capture and for fission reactions with U^{235} and U^{238} are shown in Fig. 4.2. This illustrates the earlier reasoning that U^{238} is fissionable by fast neutrons only whereas U^{235} becomes increasingly fissionable with reducing incident energy in the thermal neutron region. Also shown are three "resonant" peaks in the medium energy region, indicating the enhanced capture reaction of U^{238} within narrow ranges of incident energy.

4.3. Fission of Natural Uranium

Natural uranium consists of the three isotopes U^{234}, U^{235} and U^{238} in an abundancy ratio of approximately 0·006/0·7/99·3. Only the latter two, therefore, are significant. In the event of fission of a single nucleus within a mass of natural uranium, up to three neutrons may be ejected with energy levels ranging into the fast region. Should any of these meet the 1·5 MeV kinetic energy fission requirements of the abundant U^{238}, a further fission may result, but it is much more probable that the neutrons will be insufficiently energized, and that progressive dissipation will result from subsequent encounters with U^{238} nuclei. These neutrons tend to be captured by the abundant U^{238} when they reach the resonance region before being slowed down to the thermal region where an encounter with U^{235} would result in fission. Since, in addition, some neutrons may escape the system, and others be absorbed by impurities, a continuous chain reaction is possible only under special circumstances. These circumstances may be realized either by enrichment or by moderation. The former involves a difficult and expensive technique of isotope separation to augment the U^{235} content of the uranium, and a "fast reactor" results when this content exceeds about 50 per cent. Moderation is the more attractive technique of introducing into the uranium a moderating material which will slow down the neutrons to the thermal range as rapidly as possible, and then release them for subsequent U^{235} fission, thus avoiding U^{238} resonance capture.

Moderation is achieved by elastic collisions between the neutrons and the moderating nuclei, the mass of which should be as close as possible to that of the neutrons in order to absorb the greatest possible energy. Hydrogen, being the lightest element, is attractive as a moderator, but tends to capture neutrons excessively. It can be used in the form of water, both light (H_2O) and heavy (D_2O), when enrichment is also employed. Helium is unsuitable since it is a gas and, being inert, cannot be compounded. Of the next lightest elements, lithium and boron are excluded because of neutron capture, whereas beryllium is scarce, expensive and toxic. Carbon, in the form of pure graphite, forms a practicable moderator and can also be arranged around the reactor core to act as a reflector and minimize neutron escape. Neutron loss can also be restrained by minimizing the surface/volume ratio of the core which, in turn, implies a critical size below which the reaction chain will cease to propagate.

Control can be effected by the insertion of materials of high capture cross-section for thermal reactions, and rods of boron, cadmium, hafnium or the oxides of the rare earths are used for this purpose. The effectiveness of such control is made possible by the time delay accompanying the 0·75 per cent of neutrons which are ejected by radioactive decay following collision, as distinct from the prompt neutrons released almost instantaneously by fission.

Many different fragments may result from the fission of uranium nuclei, but the most probable products are nuclides of mass number in the region of 96 and 140, as shown in the following flow sheet for the neutron bombardment of U^{235}:

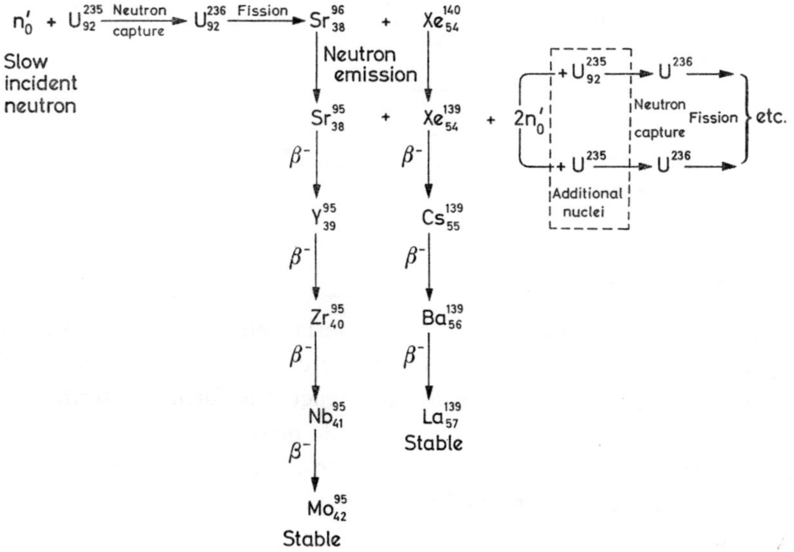

The initial products of fission are each seen to release one active neutron such that a net gain of one neutron eventuates, and chain branching is possible. The resulting fission fragments then decay by β-emission until stable nuclides are reached.

The quantity of energy evolved as a result of each fission of a U^{235} nucleus can be calculated either by reading off from Fig. 2.12 the differences in values of $\bar{B}$ for the reactants and products, or from the mass defect of the reaction, as follows:

Initial reactant mass $= M_r = 235{\cdot}117 + 1{\cdot}009 = 236{\cdot}126$ amu.

Final product mass $= M_p = 94{\cdot}942 + 138{\cdot}952 + 2{\cdot}018 = 235{\cdot}912$ amu. Thus, product mass defect $= M_r - M_p = 0{\cdot}214$ amu,

$$\equiv 0{\cdot}214 \times 931 = 199 \text{ MeV} = B,$$

and $\bar{B} = 199/236 = 0{\cdot}85$ MeV approximately. Over 80 per cent of this energy is converted into kinetic energy of fission fragments.

The isotope U^{238} also is fissionable but only by neutrons in the fast region and is not, therefore, of prime interest as a source of nuclear energy. Its neutron absorbency, particularly in the resonant range, is of practical value since the transmutation product, plutonium, is fissionable by thermal neutrons:

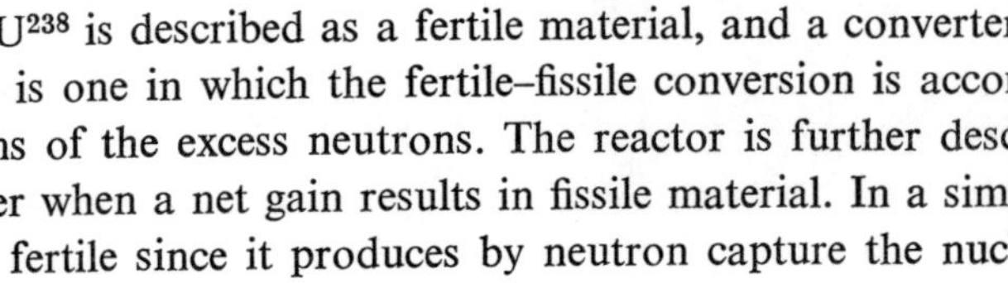

Hence, U^{238} is described as a fertile material, and a converter type of reaction is one in which the fertile–fissile conversion is accomplished by means of the excess neutrons. The reactor is further described as a breeder when a net gain results in fissile material. In a similar way, Th^{232} is fertile since it produces by neutron capture the nuclide U^{233} which is fissionable by thermal neutrons.

In summary, the following materials are of interest in the production of energy by nuclear fission:

Fertile		*Fissile*
		U^{235}
U^{238}	$\longrightarrow$	Pu^{239}*
Th^{232}	$\longrightarrow$	U^{233}*

4.4. Nuclear Fusion

From the shape of the binding energy curve in Fig. 2.12, it is evident that greater quantities of energy per nucleon are available from the fusion of light nuclei than from the fission of heavy nuclei. Since the gradient of the curve is steepest for the lightest nuclei, the hydrogen isotopes are of prime interest as fusion components, and four reactions with them are feasible:

Reaction	*B, MeV*	*$\bar{B}$, MeV*
$H_1^2+H_1^2 \longrightarrow H_1^3+H_1^1$	3·9	0·98
or		
$H_1^2+H_1^2 \longrightarrow He_2^3+n_0^1$	3·3	0·83
$H_1^2+H_1^3 \longrightarrow He_2^4+n_0^1$	17·6	3·52
$H_1^3+H_1^3 \longrightarrow He_2^4+2n_0^1$	11·3	1·88

The relative values of $\bar{B}$ are more significant in practice since these represent roughly the energy output per unit mass of nuclear fuel material.

The first two reactions are collectively described as DD (i.e. deuteron + deuteron), and the third and fourth as DT (deuteron + triton) and TT respectively. The T and He^3 resulting from the DD reaction can react with further D nuclei as in the DT reaction above, and the DHe^3 reaction below:

$$H_1^2+He_2^3 \longrightarrow He_2^4+H_1^1 \qquad 18{\cdot}4 \qquad 3{\cdot}68$$

* Not naturally occurring.

Tritium is costly to produce, whereas deuterium is naturally available in large quantities (e.g. in sea water). Since the probabilities of the two routes of the DD reaction are roughly comparable, a group of six deuterons can be considered to react as follows:

Reaction	*B, MeV*	*$\bar{B}$, MeV*
$H_1^2 + H_1^2 \rightarrow H_1^3 + H_1^1$; $H_1^3 + H_1^2 \rightarrow He_2^4 + n_0^1$		
$H_1^2 + H_1^2 \rightarrow He_2^3 + n_0^1$; $He_2^3 + H_1^2 \rightarrow He_2^4 + H_1^1$	43·2	3·6

Hence, six deuterons comprising twelve nucleons release 3·6 MeV per nucleon, the corresponding value for uranium fission being 0·85 MeV per nucleon. Roth[20] describes this as the supplemented DD reaction, and shows that over 60 per cent of the energy remains with the charged particles.

The fusion of light nuclei, however, can be achieved only at collision speeds sufficient to overcome the powerful electrostatic repulsion. This requires temperatures of about 10^{8}° K. Hence, at this stage, nuclear fission is the only practical means of realizing the necessary initial thermal energy, and the overall process is described as a thermonuclear reaction, as in the fission-fusion hydrogen bomb. For the controlled production of power, however, the magnitude of the attendant engineering problems is immense, and the most promising solution appears to be the utilization of the magnetic pinch effect in order to protect the containing walls of the reactor. Furthermore, the quantity of energy radiated is very sensitive to the presence of contaminants, and can lead to quenching. In view of the development work necessary, application of nuclear fusion to propulsion cannot be expected for some considerable time.

4.5. The Nuclear Rocket

The main objective in a nuclear rocket is to convert the energy of the reaction products into directed kinetic energy of the propellant. Reaction products comprise charged particles and neutrons, and the kinetic energy of the charged particles is available directly for conversion to thrust, whereas neutrons tend to leave the reacting region before releasing their energy. A high retention of energy in the charged particles, therefore, is a propulsion requirement. Conversion of this energy into thrust may be accomplished either indirectly by heat transfer to the propellant used as an external coolant, or directly by using the nuclear fuel as either the propellant itself or part of the propellant mixture.

4.5.1. The Heat-transfer Fission Rocket

This concept comprises a solid reactor-core matrix perforated with parallel channels for the propellant flow (Fig. 4.3). The heated propellant expands through conventional convergent–divergent nozzle geometry, cooled by the incoming propellant. The propellant may also be used to cool a reflector sheath surrounding the core. Control is effected by axially-moving rods of neutron-absorbent material located either in the core or the reflector, working on an output temperature signal.

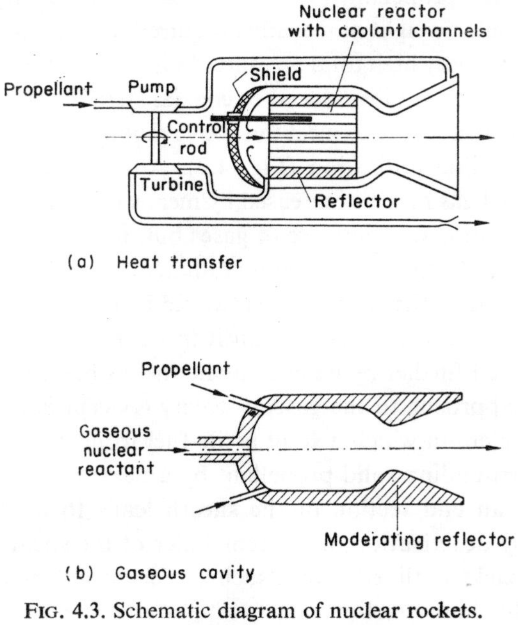

FIG. 4.3. Schematic diagram of nuclear rockets.

The dissociation of high-temperature propellant gases at very low pressures offers an increment in specific impulse due to the reduced molecular weight of the efflux, provided the reduction in thrust is acceptable. Corliss[10] shows an increase of about 40 per cent with hydrogen at a pressure of 0·01 atm compared with 1 atm. In essence, the propellant comprises a free-radical fluid which is produced as required without the handling difficulties of storage and feeding.

4.5.2. The Gaseous-cavity Fission Rocket

The intrinsic advantage of direct heating in the nuclear rocket is a higher level of performance following the easement of the metallurgical temperature limitation, and the absence of a temperature drop through imperfect heat transfer. The main drawbacks are the presence of radioactive fission products in the efflux, and the loss of only partly-consumed fissile fuel. In the gaseous-cavity rocket, the gaseous fissile fuel is introduced at high partial pressure in order to achieve criticality. However, a high propellant/fuel mixture ratio is necessary to maintain the low molecular weight of the efflux required for high specific impulse, and this infers a very high total pressure. In its simplest form, the propellant and fuel gases are introduced in coaxial streams, the latter flowing at lower velocity in order to meet the above requirements, and to achieve maximum possible utilization (Fig. 4.3). Alternatively, the residence time of the fuel gas may be increased by means of vortex flow. This provides a more intimate admixture of gases but, in each case, the propellant must be able to absorb energy radiated from the fuel fission. In a two-stage version, the propellant gas would be heated to the metallurgical-limited temperature during transit through a solid reactor matrix, and then heated further by gaseous fissile fuel as before.

An allied approach to the gaseous-cavity concept is the consumable solid-core system in which a solid fissile fuel forms the core, separated from the surrounding solid propellant by a neutron-absorbing sheath. Removal of an end section of the sheath leads to a reaction which progresses by deactivation of the remainder of the sheath. In addition to the drawbacks outlined above, such a concept suffers so far from an uncontrollably high reaction rate.

4.5.3. The Thermonuclear Rocket

The thermonuclear rocket might appear as a development of the gaseous-cavity type, with increased performance due to the greater energy release. The main problems comprise initiation of the reaction, reducing the level of energy loss by scattering and absorption below that of energy release, containment of the reaction products at exceptionally high temperature and pressure, and the virtual elimination of reaction contaminants. Advantages include the fourfold increase of energy over the fission process and, if an initiation system alternative to fission can be developed, there will be no problem of radioactive contamination of the efflux. However, the problems of containment and wall protection are severe due to the extremely high temperature in excess of 10^{8}° K. Since reacting gases will be ionized at these temperature levels, these problems might be resolved by means of the pinch effect imposed by the discharge of a heavy current through the reaction zone, and by the use of "magnetic mirror" coils surrounding each end of the zone to confine the field lines into a "magnetic bottle". By making the field strength at one end of the bottle less than that at the other, the plasma will be constrained to escape at a controlled rate (Fig. 4.4).

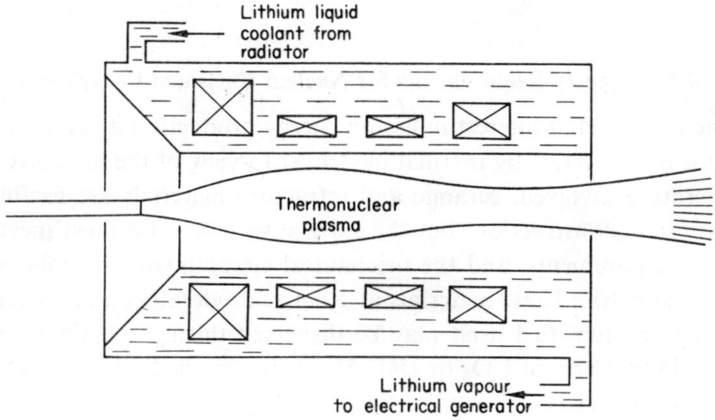

FIG. 4.4. Schematic diagram of thermonuclear rocket.

The column of plasma resulting from the pinch effect is maintained by the pressure balance between the plasma and the constraining magnetic field. The column rapidly exhibits instabilities due to kinks and constrictions, but the addition of a strong longitudinal magnetic field may provide sufficient stiffness for the column to endure. The quenching effect of the material chamber wall ensures that the reaction will not develop explosively. Weight penalties will prevail due to the heavy requirements for electricity for the containment system.

4.6. Property Requirements for Nuclear Fuels and Propellants

The main fuel of interest in fission energy production is uranium 235 since it is fissionable by thermal neutrons. In view of the high levels of temperature involved, ceramic and refractory materials are useful for reactor-core construction, but the surface/volume ratio must meet the critical requirements, and the mechanical strength withstand the operational stresses. These materials also act as moderators and reflectors, and the uranium fuel itself may be dispersed throughout the ceramic matrix in the form of UO_2 or UC. Alternatively, the fuel elements may be clad with a refractory metal as protection against corrosion or product leakage. For thermonuclear energy, deuterium offers the best combination of properties, being widely available with a high-energy release to its charged product particles, but its fusion initiation temperature is extremely high.

Since the propellant for a nuclear-fission rocket achieves its propulsive effect by thermal expansion, the property requirements are generally similar to those for a chemical rocket. However, there are two highly significant exceptions. The first relates to the fact that, in the chemical rocket, the same materials perform the twofold function of energy source and working fluid. One of the controlling requirements, therefore, concerns the energy content of the reactant, and the values of bond (i.e. chemical binding) energy for the combustion products. As shown in Chapter 2, the strength of the chemical bond is such that specific impulse is limited to less than 500 lbf. s/lbm for optimum combinations of hydrogen with such oxidants as ozone, fluorine and oxygen.

TABLE 4.1. Property Requirements of Nuclear Propellants

Property requirement / Elemental propellant	M.W. (low)	Thermal Neutron Absorption (barns) (low-fission) (high-fusion)
Hydrogen	2·02	0·33
Helium	4·00	0·0008
Lithium	6·94	71·0
Beryllium	9·01	0·005
Boron	10·82	750·0
Carbon	12·01	0·0045
Nitrogen	14·01	1·7
Oxygen	16·00	0·0006

(Derived from refs. 14, 18 and 24.)

The choice of chemical propellants, therefore, is largely restricted by combustion parameters, and these were seen to be most promising with propellant mixtures of low specific gravity.

The second exception concerns the importance of the nuclear properties of the propellant. With U^{235} as the fissile fuel, great economy of thermal neutrons was seen to be necessary for the reaction chain to progress, and a moderator is incorporated for that purpose. Hence, the propellant coming within the proximity of the fissile fuel must exhibit a low absorption cross-section for thermal neutrons.

Using low molecular weight as a prime criterion, nuclear propellants rate in descending order from hydrogen through dissociable compounds based on the first few elements of the periodic system, as shown in Table 4.1. The superimposed neutron absorption requirement is seen to eliminate both lithium and boron for fission purposes, whereas cost and handling problems render helium and beryllium unattractive also. Hence, hydrogen appears once again as the supreme element in the field of rocket propulsion.

Of the hydrides available, water is abundant and an excellent neutron moderator but is highly corrosive as high-temperature steam, and its molecular weight is high (18) since dissociation occurs only at temperatures above about 2500° K. A wide range of hydrocarbons is available, giving dissociated molecular weights of about 8 at high temperatures and pressures. Neutron moderation is good, and corrosion no problem, but fire hazards exist in storage. Similar remarks apply to methyl and ethyl alcohol. The nitrogen hydrides, ammonia and hydrazine, give dissociated molecular weights of about 10, but present health and handling hazards, and are less acceptable as neutron moderators.

In nuclear fusion, a larger fraction of the reaction energy appears as kinetic energy of neutrons, and a high absorption by the propellant is desirable. In the fusion rocket, these energetic neutrons will not be greatly influenced by any electromagnetic constraint devices, and will need to be absorbed by radiation shielding around the reaction chamber. Thermal radiation from the reaction zone will be considerable and, since the high-temperature gases will be ionized, the deceleration of charged particles in the hot plasma by the presence of electrostatic fields around heavy ions will create a kinetic-electromagnetic energy conversion resulting in an electromagnetic radiation known as "Bremsstrahlung". The shielding will therefore absorb these radiations in addition, and its resulting vaporization has been suggested as a source of propellant gas. However, quenching by propellant contamination and energy absorption must be avoided for the fusion reaction to be sustained. Lithium appears as the most promising propellant for the thermonuclear rocket, but it may have to be introduced into the fusion products at the downstream end of the reaction zone. Bremsstrahlung quenching also precludes the use of fusion rockets in an atmosphere of any appreciable pressure, since plasma contamination will occur from atmospheric ingress through the thrust orifice.

Symbols

A = Mass number
B = Binding energy
$\bar{B}$ = Binding energy per nucleon
E_A = Fission activation energy
I = Activity of radioactive disintegration
M_r = Initial reactant mass

M_p = Final product mass
N = Number of like radioactive nuclei
t = Time
$t_{\frac{1}{2}}$ = Half-life of radioactive element
Z = Atomic number
δ = Factor related to numbers of protons and neutrons
λ = Radioactive decay constant

Subscripts

o = Initial condition

CHAPTER 5

ELECTRIC PROPULSION

IN THE previous Chapter, the propulsive performance of the fission-powered rocket, in terms of specific impulse, was seen to range generally from about 1000 to 3000 lbf. s/lbm, whereas the higher performance thermonuclear rocket appears to be some way yet from feasibility. In view of thermal limitations on materials, the propellant temperatures achieved in the fission rocket are limited to about 2500° K with the solid core, and about 10,000° K with the more conceptual gaseous core. By introducing an intermediate step of thermal-electrical energy conversion, the remaining energy can be applied to the propellant in such a way to give a factor of improvement of about 8 in efflux velocity and specific impulse without additional thermal problems. Since the ratio of thrust to efflux power varies inversely with specific impulse, the thrust level of electric propulsion systems is low, of order 10 lbf if the electrical energy is converted locally back to heat, and only 1 lbf if propellant acceleration is effected by electrical forces.

The electrical service power required for spacecraft instrumentation, control, telemetry, etc., ranges from a few watts for instrumented space probes to several kilowatts in the case of manned vehicles. The power requirements for electrical propulsion systems, however, are considerably greater, and are expected to range from about 30 up to several thousand kilowatts. Of the many methods available and proposed for conversion to electrical energy, these high-energy demands are likely to be met by means of turboelectric conversion from a nuclear energy source, with the possibility of thermionic conversion given suitable development and improved conversion efficiency. For very small thrusts, the optical concentration of solar radiation may suffice as an energy source, particularly for missions in the vicinity of the Sun. The turbo-electric system will probably operate on the Rankine thermodynamic vapour cycle, using potassium or other alkali metal in a closed system. Two-phase vapour cycles permit heat addition and rejection at the highest and lowest cycle temperatures respectively, but the overall cycle efficiency may not exceed about 30 per cent. A reduction in heat-rejection temperature increases cycle efficiency, but also increases the size of the waste heat radiator and its vulnerability to meteoroid damage.

Direct thermionic conversion of heat to electrical energy operates through the release of electrons from the heated surface of a solid cathode. Present conversion efficiencies appear to peak at about 15 per cent, but improvements are likely with continued development. Again, solar energy may prove an acceptable source of thrust power, coupled with direct conversion by means of solar cells.

5.1. Plasma

The simple representation of a fluid such as a monatomic gas is a collection of minute spherical particles which attract each other gravitationally at remote spacing, and repel electrostatically at close spacing. The spheres move randomly in straight lines until directions are changed by collision. Clearly, collision will ensue only when spacing between the centres of the two spheres attempting to pass each other is less than the combined radii of the spheres, i.e. the centre of one sphere of radius r_1 must strike within the projected circle drawn at radius (r_1+r_2) from the centre of the second sphere of radius r_2. The projected target area of $\pi(r_1+r_2)^2$ is termed the collision cross-section. In fact, atomic particles do not comprise hard spheres, and the effective collision cross-section is a function both of the dimensions of the particle and its ability to disrupt the target particle. Hence, the collision cross-section of electrons is high, despite their small magnitude, due to their mobility and rapid motion.

The continued addition of energy to a fluid gives rise progressively to vaporization (if liquid), dissociation (if polyatomic), and eventually ionization to produce free electrons and parent ions. All gases possess some ionized particles, but a gas with a relatively high degree of ionization is termed a plasma, and is produced by the addition of a substantial quantity of "ionization" energy. The energy may be added thermally by high-temperature combustion, heat transfer or shock waves, or electrically by heavy discharge of current. As shown in Fig. 2.9, however, the magnitude of ionization energy is substantial relative to conventional combustion energy hence, in the early stages of development, systems employing plasmas are likely to be dependent upon a nuclear energy source.

Further energy addition increases ionization both in extent and in multiplicity, with the involvement of more than one electron per atom. The resulting free electrons and ions each possess an electrical charge and are therefore capable of imparting electrical conductivity. In a macroscopic quantity of fluid, ionization and recombination processes will be occurring simultaneously, and an equilibrium degree of ionization will exist depending upon the mean level of energy of the fluid. The equilibrium temperature of the plasma is related to particle velocity by the expression

$$T = mv_r^2/3\,k_B,$$

where m = mass of particle, v_r = root-mean-square velocity of particle, and k_B = Boltzmann's constant = $1{\cdot}3804\times10^{-23}$ joule/atom °K. Hence, mean energy level = $\frac{1}{2}mv_r^2 = 3k_B T/2$, which gives the energy–temperature relationship of 1 eV ≡ 7740° K approximately. The high-energy content of plasma may be converted directly to kinetic energy by the conventional process of expansion in a convergent–divergent nozzle, but the electrical properties of plasma permit further acceleration by means of added electrical energy.

Overall, a plasma is electrically neutral, but possesses an electrical conductivity which results largely from the free electrons which are light and mobile compared with the parent ions. The electrons and ions will remain in equilibrium quantities until either the added energy is dissipated, or the electrons and ions are separated into independent groups. In the event of a propellant plasma being separated in this way, acceleration of the charged particles of one sign only can be imparted by the application of an axial electrostatic field. Separation of electrons from a plasma stream is possible by means of an electron-absorbing medium or an electromagnetic field, and the resulting ionized fluid can be utilized in a jet device. The separated electrons must also be ejected in order to avoid the accumulation of a negative space charge on the vehicle, but the particle masses are such that the additional thrust is negligible.

Since electromagnetic fields of opposing direction repel each other, parallel conductors carrying current flowing in opposite directions also repel mutually. If one conductor takes the form of a plasma, it will be free to move under repulsion so long as the current continues to flow.

TABLE 5.1. Systems of Electric Propulsion

Propellant fluid	Accelerating device	System
Current-maintained plasma flow	Heater chamber and convergent-divergent nozzle	Electrothermal rocket
Plasma flow with free electrons removed	Electrostatic field	Ion jet, or electrostatic rocket
Plasma in discrete accumulations (plasmoids)	Electromagnetic field	Plasmoid accelerator
Continuous plasma flow	Crossed electrostatic and magnetic fields	MHD rocket

When plasma flows continuously without ion separation, approximately equal quantities of both electrons and ions are involved. In order to produce acceleration of both types of particle in the same direction, both magnetic and electrostatic fields must be applied, mutually perpendicular to the direction of flow. Four methods exist, therefore, for the application of electrical energy to rocket propulsion, as shown in Table 5.1, and these systems are discussed in further detail in the following sections.

5.2. Electrothermal Propulsion

The design of these systems follows that of the conventional reaction chamber plus convergent–divergent nozzle, with chemical combustion replaced by either an arc discharge or a resistance-heated exchanger located concentrically within the chamber, known respectively as an arcjet and a resistojet. In the former case, the propellant is introduced vortex-wise in order to shield the walls from the high arc temperatures. In approaching and traversing the arc, the propellant is converted progressively into a plasma, as discussed above. The resultant cooling of the outer sheath of the arc promotes a thermal pinch effect constricting the current within the core and raising the operating temperature to very high values of order 50,000° K (Fig. 5.1).

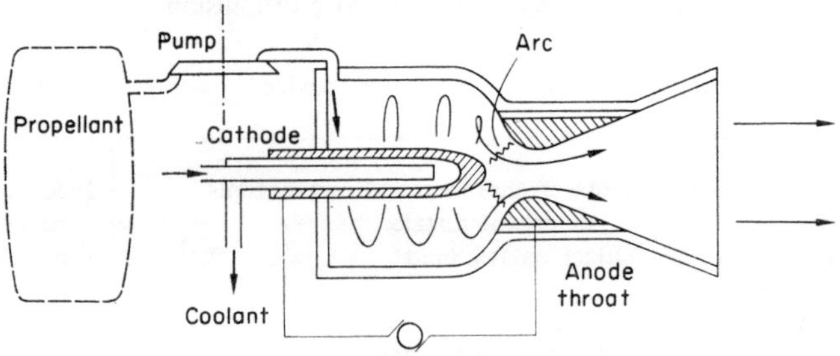

FIG. 5.1. Schematic diagram of arc-heated electrothermal rocket.

This addition of energy to the propellant, and its resultant thrust performance, is not restricted by the atomic bond energies of chemical combustion. Nevertheless, the energy absorbed in dissociation and ionization of the propellant is usable only if it can be regained during the supersonic diffusion. High operating pressures are helpful in suppressing dissociation and ionization, but the very high efflux velocities, of approximately 50,000 ft/s, represent extremely short residence times. For this reason, frozen-flow conditions are assumed, and the peak specific impulse obtainable is limited to about 2000 lbf.s/lbm due to nonrecombination and to radiation. Practical problems also ensue through electrode erosion, and the massive requirements for electrical power. The expected thrust/power ratios at high specific impulse are of order 0·12 lbf/kW on the basis of efflux power, and about one third of this on the basis of overall power consumed. This represents a thrust of about 1 lbf for an input rate of 30 kW.

Specific impulse increases rapidly with added enthalpy until dissociation and/or ionization commence. Here, the rates of increase are more gradual due to the enthalpy absorbed in these partition processes, and the levels of frozen-flow efficiency show corresponding depressions. When ionization is complete, specific impulse again increases rapidly with added enthalpy, and frozen-flow efficiency shows a tendency to increase (see Fig. 5.13 in Section 5.5).

5.3. Electrostatic Propulsion

The four processes basic to electrostatic propulsion comprise the ionization of neutral material, the separation of electrons and ions, the acceleration of ions, and the subsequent recombination of electrons and ions to produce a neutral propelling jet. As an introduction, the fundamental properties of electrostatic fields are reviewed, together with the forces experienced by charges located within them.

5.3.1. Electrostatic Fields

An electrostatic field is the volume of influence surrounding and resulting from an electric charge. The charge itself may be produced by rubbing dissimilar materials together, or by transferring a quantity of electrical energy to a body with capacity to receive it. The charge exists because of either a deficiency or an excess of electrons causing an electrical unbalance in the body.

The quantity of electrical energy comprising a charge is equivalent to a flow of current for a specified period, and the standard unit of charge is the ampere-second or coulomb. Ideally, the electrostatic field generated by an electric charge is visualized concentrated at a point with a series of flux lines radiating isotropically from it. The sign of electric charges was determined arbitrarily in the eighteenth century, and the field concept depends upon an outwards-directed flux for a positive charge, and an inwards flux for a negative charge. The direction of the flux is defined as the direction of the force which would be exerted on a positive test charge located within the field of the source charge under consideration (assuming the test charge does not disturb the field), hence the repulsion arising in the case of the positive force field.

In any sphere concentric with the source charge, the total flux through the surface of the sphere is equal to the source charge itself. Using the following symbols, together with S.I. units:

q_1 = strength of source charge, C,
r = radius of concentric sphere, m,
D = flux density = flux per unit area, C/m², the total flux through the sphere surface $= q_1 = D(4\pi r^2)$. Hence, $D = q_1/4\pi r^2$, C/m². D, as written, represents the magnitude only of a vector quantity which has both magnitude and direction. In vector notation, this symbol is printed in bold-face type, **D**.

When a test charge q is introduced into the stationary field at the surface of the sphere, the force exerted upon it is proportional directly to the product of the charge strengths, and inversely to the square of the charge separation distance, acting in the direction of the line common to both charges, i.e.

F_s = electrostatic force on test charge $\propto q_1 q/r^2$ (Coulomb's law),

as shown in Fig. 5.2. Since the test charge is located in a flux density D, it is equally true to state that the force is directly proportional to the product of the test charge strength and the flux density:

$$F_s \propto qD = kqD = kq_1q/4\pi r^2.$$

In a vacuum, the proportionality factor k is expressed as $1/\varepsilon_0$, where ε_0 is known as the dielectric constant of a vacuum, or the electric permittivity of free space, and is equal to $8{\cdot}854\times 10^{-12}$ farad/metre.†

† For clarity, the names of the units have been written in full for some of the equations which follow.

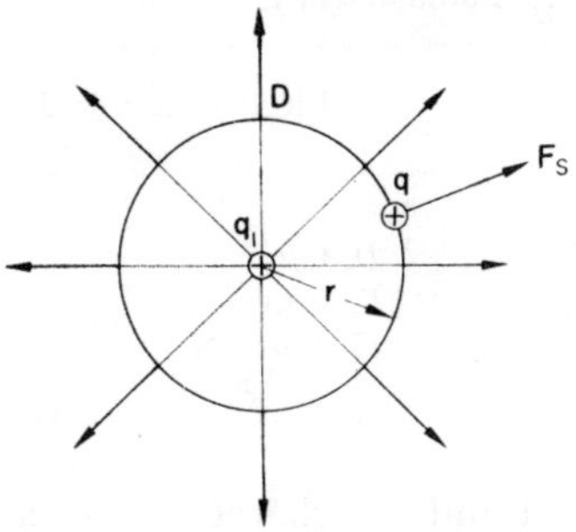

FIG. 5.2. Electrostatic field and force.

Hence, $F_s = q_1q/4\pi\varepsilon_0 r^2$, newton.

Since $F_s = qD/\varepsilon_0$, the force exerted on a unit test charge is given by $F_s/q = D/\varepsilon_0 =$ Electric field intensity $= E = q_1/4\pi\varepsilon_0 r^2$ newton/coulomb (= volt/metre).

Hence, $F_s = qE$, newton.

In vector notation, $\mathbf{F}_s = q\mathbf{E}$, newton. (5.1)

In any material field, $k = 1/\varepsilon\varepsilon_0$, where ε is the specific dielectric constant of the material, and is non-dimensional. Hence, the flux density, or number of flux lines per unit area, is not influenced by the presence of the material, whereas the field intensity, or strength of the lines, is attenuated by it.

If the positive test charge located in a positive source field is displaced towards the source charge, work done on the test charge in opposing the repulsion appears as a gain in potential energy by the test charge, i.e. Work done $= dW = -\mathbf{F}_s \cdot \mathbf{dr} =$ potential energy gain, newton. metre (= joule).

The total potential energy gained by the test charge in moving into its position from infinity

$$= W = -\int_{\infty} \mathbf{F}_s \cdot \mathbf{dr}.$$

TABLE 5.2. Summary of Electrical Parameters

Parameter	Electrostatic field	Magnetic field
Flux density	D, C/m^2	B, Wb/m^2
Field intensity	E, V/m	H, A.turn/m
Force on q in vacua	$F_s = qE$, N	$F_M = qvB$, N

Expressed on a basis of unit test charge, this becomes

$$W/q = \text{Electric potential} = V = -\int_{\infty}^{r} \frac{\mathbf{F}_s}{q} \cdot d\mathbf{r} = -\int_{\infty}^{r} \mathbf{E} \cdot d\mathbf{r} = q/4\pi\varepsilon_0 r$$

joule/coulomb (=volt).

Hence all test charges in a given spherical surface concentric with the source charge have the same potential, and the surface is described as equipotential. A potential difference exists between two such surfaces.

5.3.2. Production of Ions

As noted earlier, substantial quantities of energy are necessary for ionization, and these may be supplied thermally, by electric arc, by radio-frequency induction heating, by contact potential, or by electron bombardment. In electrostatic propulsion, the objective is to produce separated ions in quantity, and two systems of current experimental interest are those of contact potential and electron bombardment, sometimes described respectively as surface ionization and volume ionization. Both systems depend upon the emission of charged particles from heated surfaces.

Application of sufficient heat to a suitable cathode located in a vacuum permits emission of electrons energetic enough to traverse a gap, collect upon a relatively cool anode, and return to the cathode through an external circuit. The potential difference developed across any external load also acts between the cathode and anode producing a space charge effect which opposes the flow of electrons. Hence, only the most energetic electrons reach the anode, the less energetic forming an electron cloud within the gap as they change direction and return to the cathode.

When an electron gains sufficient energy to escape from the solid surface of the cathode, an image charge to opposite sign may be imagined induced in the cathode at a depth from the surface equal to the displacement of the escaped electron ($= x$ in Fig. 5.3). From Section

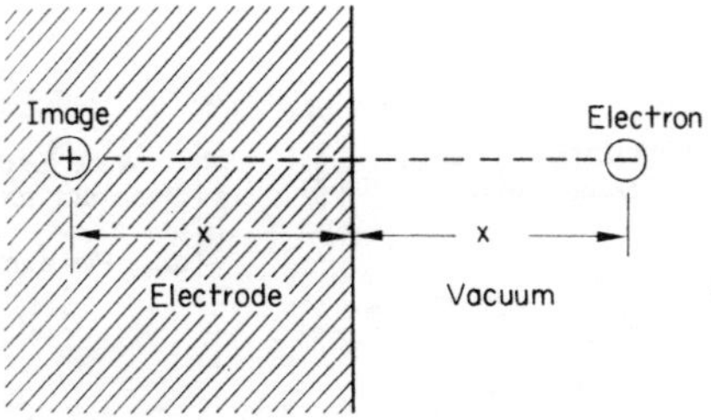

FIG. 5.3. Image charge resulting from surface-released electron.

5.3.1 the surface-escape of a charge q into a vacuum gives rise to the following:

$$D = q/4\pi r^2,$$
$$E = q/4\pi\varepsilon_0 r^2,$$
and
$$F = q^2/4\pi\varepsilon_0 r^2 = q^2/16\pi\varepsilon_0 x^2, \text{ since } r = 2x.$$

Thus, work required to free electron from surface

$$= \int_{x_1}^{\infty} F\,dx = q^2/16\pi\varepsilon_0 x_1 = q\varnothing \text{ joule,}$$

where x_1 is the minimum value of x over which F holds, and is of order of atomic spacing. The surface work, or work function, $\varnothing$ is the potential equivalent to the work required to free an electron from the cathode surface. Values range from about 1·8 volt for caesium to 5·3 volt for platinum, and are shown in Fig. 5.4.

A special case arises when a propellant fluid is introduced into the inter-electrode gap, and the energy supplied is slightly less than the work-function level of the cathode material, but in excess of the first ionization level of the fluid. The cathode surface is unable to free its own electrons but will continuously adsorb the atoms of the fluid, strip off the outermost electrons, and release the resulting ions. Some values of first ionization potential are included in Fig. 5.4, and this begins to indicate possible surface-propellant pairs for "contact potential" operation.

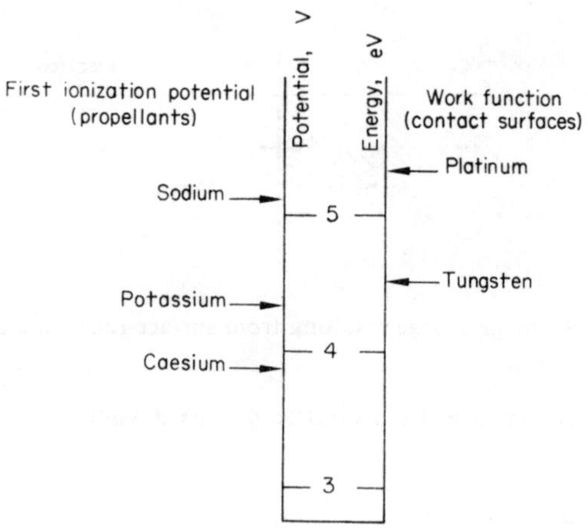

FIG. 5.4. Relationship between propellants of low ionization-potential and surfaces of high work function.

In the electron-bombardment system, electrons are discharged from a heated cathode, which may be located on the axis of a cylindrical anode in order to ionize by transverse bombardment the neutral particles flowing coaxially through the anode duct. Flow through an anode screen then traps the freed electrons and permits the ions to continue to the accelerating section.

5.3.3. *Acceleration of Ions*

The electrostatic acceleration system for propellant ions may be considered as a special case of a capacitor in which the anodic charges become free to move.

The total potential energy in an electrostatic field surrounding a charge may be found by integration throughout all space within the field, to give:

$$W_s = \tfrac{1}{2} \int ED \, d\tau, \text{ joule,}$$

where $d\tau$ is a volume element.
For the electrostatic field in the gap between the parallel plates of a capacitor, work required to increase the plate separation represents a change in the total potential energy of the field, i.e.

$$dW_s = (\tfrac{1}{2} ED)\, A \, dl = F_s \, dl,$$

where A is the plate area and dl is the change in separation distance.

$$\text{Hence, electrostatic pressure} = F_s/A = \frac{ED}{2} = \frac{\varepsilon_0 E^2}{2} = \frac{\varepsilon^2}{2}\left(\frac{\Delta V}{l}\right)^2$$

$$= 4{\cdot}427 \times 10^{-12}\left(\frac{\Delta V}{l}\right)^2 \text{ N/m}^2. \qquad (5.2)$$

with ΔV, the potential difference between the plates, in volts and l in metres. Pictorially, an annular design of parallel-plate capacitor may be represented as in Fig. 5.5A showing the tension resulting from the electrostatic flux lines extending from the cathode to the anode.

When an anodic charge is free to move under the action of applied forces, the tension acts between the cathode and the charge, as in Fig. 5.5B, and the charge accelerates towards the cathode aperture in much the same way as an arrow accelerated by the elastic tension imparted through the bowstring. (In the figure, the charge is assumed sufficient in magnitude to warrant drawing two flux lines in order to illustrate more clearly the axial direction of the resultant acceleration.) The presence of the free charge within the gap will reduce the voltage gradient between the anode and the charge, and correspondingly increase the gradient between the charge and the cathode. With sufficient charges, as in Fig. 5.5C, a condition is reached where all flux lines terminate on charges, the anode charge layer is effectively drawn a short distance closer to the cathode, and the anode itself is ineffective. This condition is described as space-charge limited, and represents a zero voltage gradient

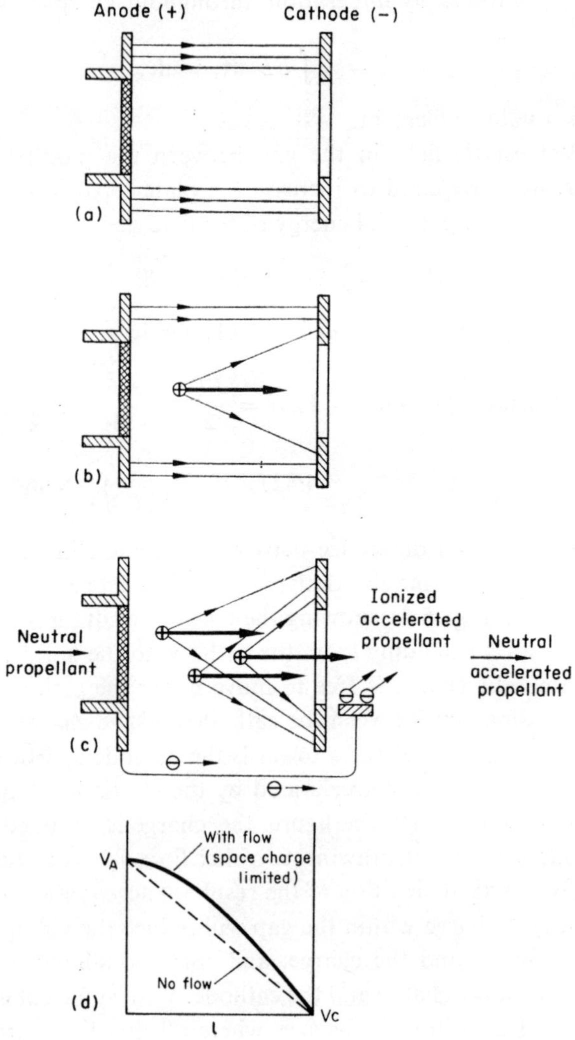

FIG. 5.5. Electrostatic acceleration of ions.

at the anode, with an increased gradient at the cathode (Fig. 5.5D). In practice, operation below the space-charge limit may be desirable in order to provide some initial acceleration for the newly created ions and thus assist in their formation. The equilibrium distribution of charges within the gap is represented by a space-charge equation known as Child's law, i.e.

Current density = charge flow rate per unit area

$$= J = \frac{4}{9}\varepsilon_0 \sqrt{\frac{2q}{m}} \frac{\Delta V^{3/2}}{l^2}\ \text{A/m}^2,$$

where $\boldsymbol{m}$ = mass of charged particle, and q = charge of particle. Since force equals rate of change of momentum,

$$\frac{F_s}{A} = \frac{\dot{m}v_e}{A} = \frac{(Jm)}{q} v_e,$$

where $\dot{m}$ is the mass flow rate of charged particles (comprising the efflux of an electrostatic rocket).

Also, from conservation of energy;

Electrical energy consumed = kinetic energy generated.

$$\text{Thus,}\quad q\Delta V = \tfrac{1}{2} m v_e^2, \tag{5.3}$$

$$\text{and}\quad v_e = I_{sp} = \sqrt{2q\Delta V/m}. \tag{5.4}$$

$$\text{Hence,}\quad \frac{F_s}{A} = \left(\frac{m}{q}\right) J v_e = \left(\frac{m}{q}\right)\left(\frac{4}{9}\varepsilon_0 \sqrt{\frac{2q}{m}} \frac{\Delta V^{3/2}}{l^2}\right)\sqrt{\frac{2q\Delta V}{m}}$$

$$= \frac{8}{9}\varepsilon_0 \left(\frac{\Delta V}{l}\right)^2 = 1\cdot\dot{7}\,\frac{\varepsilon_0}{2}\left(\frac{\Delta V}{l}\right)^2 = 7\cdot 86\times 10^{-12}\left(\frac{\Delta V}{l}\right)^2\ \text{N/m}^2. \tag{5.5}$$

In comparison with equation (5.2) for the no-flow parallel plate capacitor, therefore, the gradient of the $\Delta V/l$ curve at the cathode is $1\cdot\dot{3}$ times, and the accelerating force acting between the charges and the cathode is $1\cdot\dot{7}$ times. A system of mass ejection thus eventuates, with a resulting forward thrust acting upon the accelerating cathodic component of the system.

Calculated values of specific impulse are plotted against voltage difference in fig. 5.6, and these indicate an attractive propellant economy. Included in the Figure are values of electrostatic pressure plotted against voltage gradient. These values show that thrust per unit area, and consequently thrust, are of a very low order, so that continuous operation over an extended period of time may be necessary in practice.

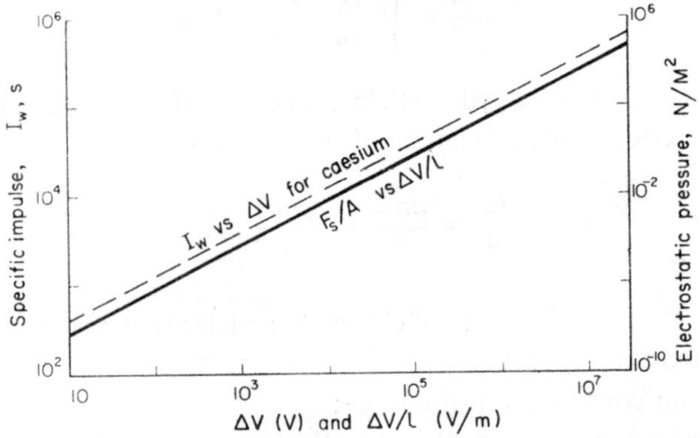

FIG. 5.6. Variation of specific impulse with voltage difference, and electrostatic pressure with voltage gradient. Reprinted by permission of the publisher, from Donald L. Turcotte, *Space Propulsion* (Waltham, Massachusetts: Blaisdell Publishing Company, 1965) (ref. 25).

Thrust may be derived from the electrostatic pressure equation as

$$F_s \propto \Delta V^2 A/l^2 \propto \Delta V^2 (D/l)^2 \propto \Delta V^2 R^2, \tag{5.6}$$

where R, the ratio of the emitter diameter to the electrode spacing, is known as the aspect ratio of the ejected beam. Improved thrust demands higher values of aspect ratio but the charges, having traversed the aperture in the cathode plane, will be decelerated to a standstill in a turnaround or stalling distance approximately equal to the spacing of the electrodes, or accelerating distance, in a manner analogous to an arrow unable to release itself from its bowstring. In the electrostatic case, this release is achieved by the use of narrow beams of charges, and by neutralizing each charge with an electron immediately downstream of the cathode. In this way, the electrostatic flux lines disengage and apply themselves to fresh anodic charges.

The practical value of aspect ratio is limited to a maximum of about unity, hence multi-parallel beams are proposed, achieved by means of a grid construction for the accelerating cathode. The close proximity of the beams also minimizes the overall heat losses from the ion emitters.

Effective neutralization demands that the net efflux current be zero, hence the numbers of electrons and ions emitted must match. Furthermore, each charge within the entire volume of the ion beam must be individually neutralized, and analysis shows that the velocities of the electrons and ions should also be closely matched. Viewed in isolation, ejection from a space vehicle can be continuous only if the efflux is electrically neutral such that the balance of the system potential is maintained, otherwise an overall space charge would arise, attracting the efflux back to the vehicle.

The simplest way to provide free electrons for neutralization will probably be by means of an annular thermionic source surrounding the beam, but the beam will lose energy in attracting the electrons, and velocity-matching may be difficult. With a high-density beam, the electrons may require an initial injection velocity. Early neutralization also minimizes the defocusing effect of the mutual repulsion forces between the charges.

For compactness in engine design, maximum utilization of electrode area is desirable, and thus a high-voltage gradient in the electrode gap. The limit is set by arc breakdown in the gap, but this is such to require massive power plants of relatively short operational life. In practice, a reduction in the resultant level of specific impulse is tolerable, and a common proposal is to accelerate the charges beyond the required velocity in order to minimize electrode area, followed by deceleration as necessary. This process also prevents the loss of neutralizing electrons from attraction upstream by the ion-emitting anode. The theoretical requirement for minimum electrode spacing also aids compactness, and values of 1 to 3 mm are typical, although practical difficulties arise in manufacture and in operation.

The termination of accelerated ions on the cathode grid leads to heating and surface sputtering, both of which can severely impair the efficiency and reliability of the unit, and limit its operational life. The eroded material does not contribute to thrust, and a balance must be struck between engine life and efficiency. Some avoidance of impingement is provided through suitable focusing of the ions by electrostatic, and possibly magnetic, means.

Figure 5.7 is a schematic diagram of an electrostatic rocket showing the two main systems of ion production discussed above, together with the systems of acceleration-deceleration, and electron neutralization.

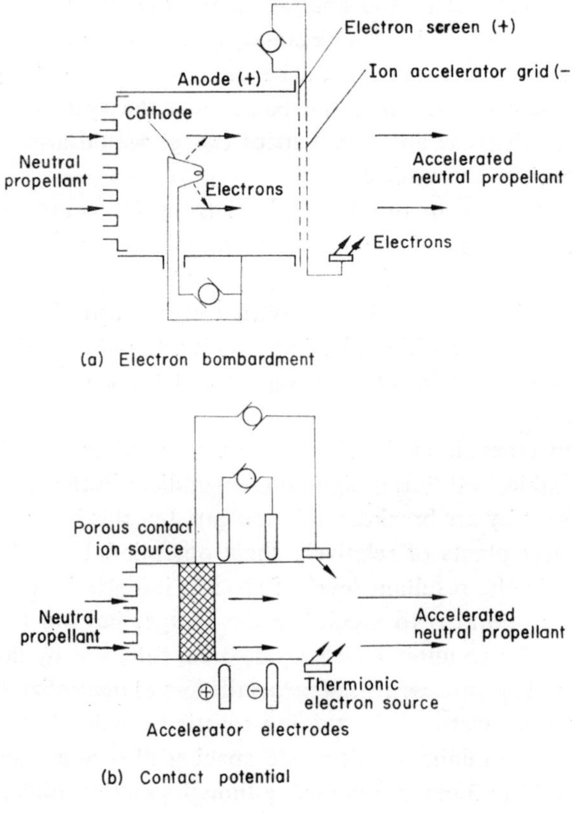

FIG. 5.7. Schematic diagram of electrostatic rockets.

5.4. Electromagnetic Propulsion

5.4.1. Magnetic Fields

A magnetic field is the volume of influence, detectable by either a test magnet or a moving test charge, surrounding and resulting from either a source magnet or a moving electric source charge. The electrostatic field associated with the source charge, discussed above, applies irrespective of movement of the charge (provided the velocity does not approach the speed of light), and the electrostatic field equals qE as before. Movement of the source charge, however, produces an additional, magnetic, field, and any movement of the test charge perpendicular to this field gives rise to an additional, magnetic, force directly proportional to the intensity E and velocity u of the electrostatic field, together with the strength q and velocity v of the test charge itself. Allowing for appropriate components of velocity relative to the plane of interest, therefore,

$$\text{Magnetic force} = F_M = k(Eu)(qv), \quad \text{N.}$$

From experiment, the constant of proportionality is found to be $1/c^2$ where c is the velocity of light $= 2{\cdot}998 \times 10^8$ m/s, thus,

$$F_M = qv(Eu)/c^2 = qvB, \quad \text{N},$$

where $B = Eu/c^2$ = magnetic flux density, V.s/m^2, or weber/m^2. Hence, the total force acting on the test charge q moving at perpendicular velocity v in a moving electric field of electrostatic field intensity E and magnetic flux density B is given by

$$\mathbf{F} = \mathbf{F}_s + \mathbf{F}_M = q(\mathbf{E} + \mathbf{v} \times B), \quad \text{N.}$$

This is shown in vectorial notation since F_s acts in the direction of the electric field whereas F_M acts normally to both the test charge velocity and the magnetic field (Fig. 5.8). The direction of B resulting from axial movement of a positive current is determined by the right-hand corkscrew rule, illustrated in Fig. 5.8A.

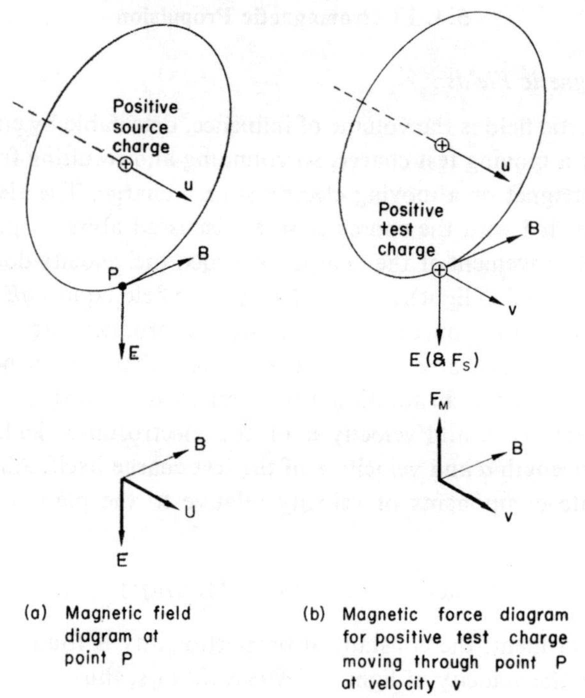

FIG. 5.8. Electromagnetic field and force.

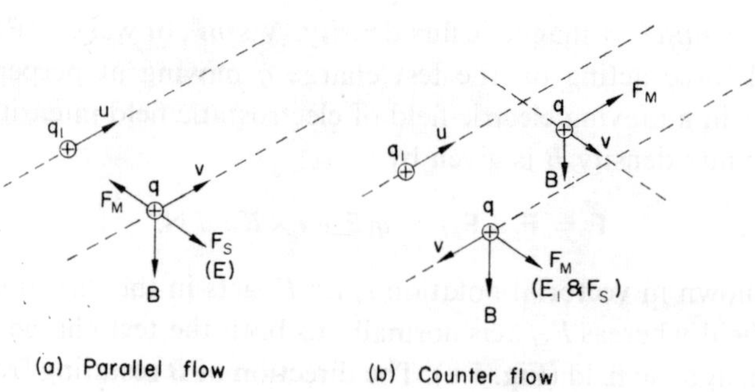

FIG. 5.9. Charges in parallel motion.

An application of the magnetic field and force diagrams is given in Fig. 5.9 for the case of parallel velocities of charges found in a parallel conductor arrangement. As shown, the charges are repelled by the electrostatic force and attracted by the magnetic force. When the charge flows are in opposing directions, both F_s and F_M act as repelling forces. Furthermore, any charges flowing across the gap from one conductor to the other will experience electromagnetic force F_M acting in the same direction as the charge flow in the donor conductor.

As shown above, $B = Eu/c^2 = \dfrac{q_1}{4\pi\varepsilon_0 r^2}\dfrac{u}{c^2}.$

Replacing $1/\varepsilon_0 c^2$ by μ_0, known as the permeability of a vacuum, equal to 4×10^{-7} henry/m, gives

$B/\mu_0 = q_1 u/4\pi r^2 = H$, A.turn/m, the intensity of the magnetic field at radius r produced by charge q_1 moving at velocity u m/s. Electrostatic and magnetic parameters are summarized in Table 5.2.

Since $q = \text{current}\times\text{time} = it$, and velocity = distance/time $= l/t$,

$$F_M = Bvq = B(l/t)(it) = Bli,$$

and $F_M/l = Bi$, N/m = magnetic force per unit length of conductor carrying current i in magnetic field of flux density B.

5.4.2. Electromagnetic Acceleration of Plasmoids

Electromagnetic forces may be applied to the acceleration of plasma in a variety of ways, including the systems of acceleration of discrete parcels of plasma, described as plasmoids. In the Bostick gun[6], a pulse of high voltage applied between the ends of two conductors causes an arc discharge. As shown in the previous section, the current-carrying plasma within the arc experiences an electromagnetic force repelling it from the end of the supply conductor. As the arc grows into a toroidal shape, differences between magnetic pressures inside and outside the toroid assist in the acceleration. Since the arc is sufficiently energetic to vaporize metallic atoms from the conductors, additives of desired molecular weight can be incorporated into the conductor material.

The mutual repulsive force exerted between parallel conductors with counter-flow of current can also be harnessed for propulsive purposes by arranging that one conductor takes the form of a plasma which can be subjected to the accelerating force for a short but continuous period of time. This leads to the proposal for a rail accelerator (Fig. 5.10) which assists the formation of a plasma discharge by an initial reduction of the gap, and permits the plasmoid so formed to continue drawing energy from the system as it accelerates along the rails away from the parallel conductor forming part of its own circuit. Acceleration forces act both from the parallel dispositions of the plasmoid with the supply

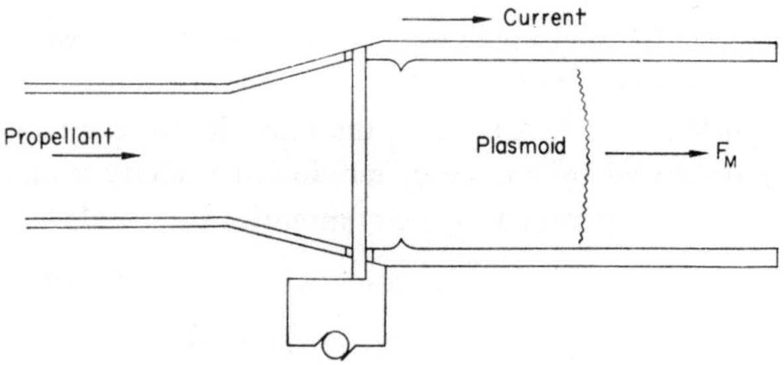

FIG. 5.10. Schematic diagram of electromagnetic rail accelerator.

conductor, and from the perpendicular disposition of the plasmoid with the two rails. An soon as the plasmoid detaches itself from the end of the accelerator, a further quantity of electrical energy can be supplied, together with additional feed gas to provide the plasma material, and the system continues in operation on a pulsed basis. For convenience and greater effectiveness, the rail-type accelerator may be arranged in coaxial form. Several other geometrical configurations have been proposed for pulsed electromagnetic acceleration, including axial coil systems producing travelling wave acceleration for the plasmoids.

5.4.3. Crossed Electrostatic and Magnetic Fields

The acceleration of continuously-flowing plasma can be achieved by means of crossed electrostatic and magnetic fields. Figure 5.11A shows two test charges of opposite sign moving at equal velocity between two parallel plates and subjected to an externally-applied magnetic field parallel to the plates and perpendicular to the flow velocity. Application of Fig. 5.8B shows a vertical magnetic force directing the positive test charge towards the upper plate (Fig. 5.11B). A corresponding diagram for the negative charge would indicate a vertical magnetic force towards the lower plate. The upper and lower plates thus become charged positively and negatively respectively, and a current flows

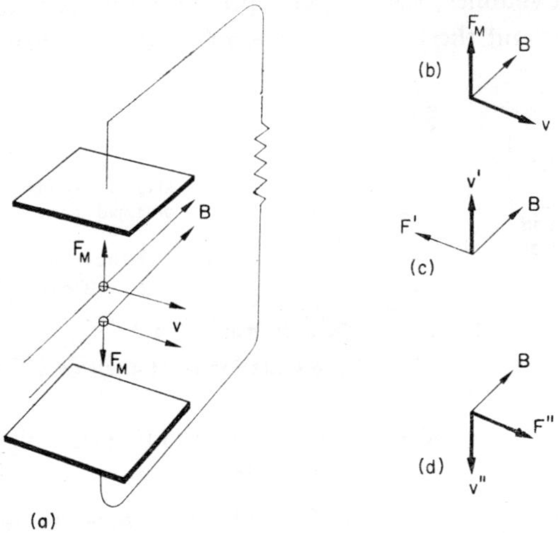

FIG. 5.11. Test charges flowing across magnetic field.

through the circuit comprising the external load. The upwards drift of positive charges at velocity v' produces an additional magnetic force F' in opposition to the original charge velocity v (Fig. 5.11C). A similar opposing force results for the negative test charge. As shown earlier, $F' = Bli$, where l is the distance between the plates, and i is the current resulting from the drift of the charges. Hence, mechanical energy extracted from the charge flow $= F'v = (Bli)\,v = (Blv)\,i = E_i i$, where $E_i = (Blv) =$ internal e.m.f. of the mechanical–electrical conversion system.

Replacement of the external load by a source of e.m.f. opposing and exceeding the internal value E_i results in a reverse drift of the test charges at velocity v'', together with a reverse force F'' accelerating the charges in the direction of the original stream velocity (Fig. 5.11D). Thus, efflux acceleration results from the mutually perpendicular application of electrostatic and magnetic forces.

5.4.4. *Magnetohydrodynamics*

The above processes for individual test charges could be described as magnetodynamics, the former representing a method of generation of electricity and the latter a means of producing thrust. In order to

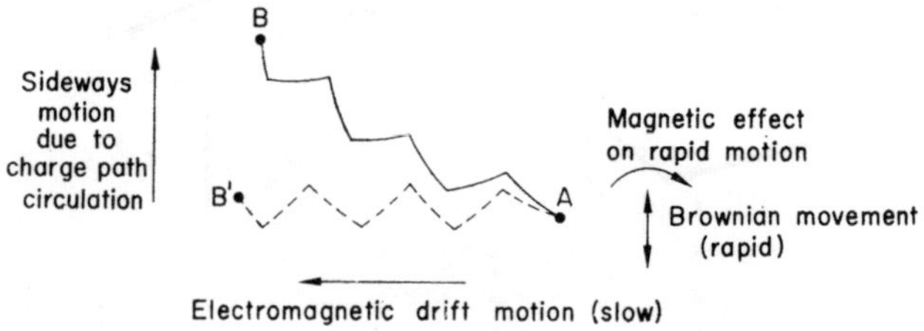

FIG. 5.12. Representation of Hall effect.

achieve practical quantities of electricity and thrust, a high conductivity is necessary, and therefore a high density of charges in the working fluid. Such a fluid is subject to the laws of hydrodynamics in addition, and the combination of the two sets of laws constitute the study of magnetohydrodynamics (MHD).

The need for plentiful free electrons represents both a substantial density of working fluid and a relatively substantial degree of ionization. The concentration of free electrons, therefore, is increased by seeding (see Section 5.5) and by maintaining a substantial fluid pressure.

The latter technique is necessary also in view of the circular nature of the thermal movement of electrons in the presence of a strong magnetic field. This follows from the fundamental effect of a magnetic field upon charge velocity, discussed above, which superimposes a constant clockwise (or anticlockwise depending upon the direction of B and the sign of the charge) rotation tendency upon the relatively rapid Brownian motion due to thermal agitation. The slower overall drift motion resulting from the action of electromagnetic forces on the main flow velocity, therefore, is biassed to one side. This is known as the Hall effect (Fig. 5.12), and any excessive mean free path for the electrons would accentuate the sideways motion and reduce severely the conductivity of the fluid in the direction of the electrostatic field. A minimum practicable fluid pressure arises, therefore, in order to minimize the mean free path. This in turn reduces the effectiveness of ionization by collision, hence the ionization process must be carried out upstream.

A fluid with a high value of electrical conductivity confines the magnetic field within the fluid, and a powerful magnetic field will then exert a very high pressure on the fluid. The magnetic field may be self-induced by means of a high current discharge through the fluid, and the resulting pinch effect constricts the fluid away from the container walls. This effect offers promise towards the solution of thermal problems with constructional materials.

As indicated earlier, theoretical considerations of MHD comprise the laws of both magnetodynamics and fluid flow. A rigorous treatment of the subject for the simple parallel duct of Fig. 5.11 includes the conservation of mass, momentum and energy together with the magnetic force discussed in the previous section, and the usual thermodynamic relationships. Such a treatment shows that the conversion effectiveness becomes progressively lessened along the duct length unless the cross-section of the duct is increased to form a divergent nozzle.

5.5. Property Requirements for Electric Propellants

As before, the requirements for propellant properties depend upon the desired performance and the type of mission, but some general conclusions can be drawn. Since all systems of electrical propulsion depend upon the production of plasma, ease of ionization forms the first requirement, and Fig. 2.9 shows a high molecular weight to be attractive. The electrons most prone to be freed from the parent atom are those in outermost orbits, particularly those which are in sole occupancy, as in the alkali metals, and marked decreases in ionization energy are indicated for these materials in Fig. 2.9. Addition of the ionization

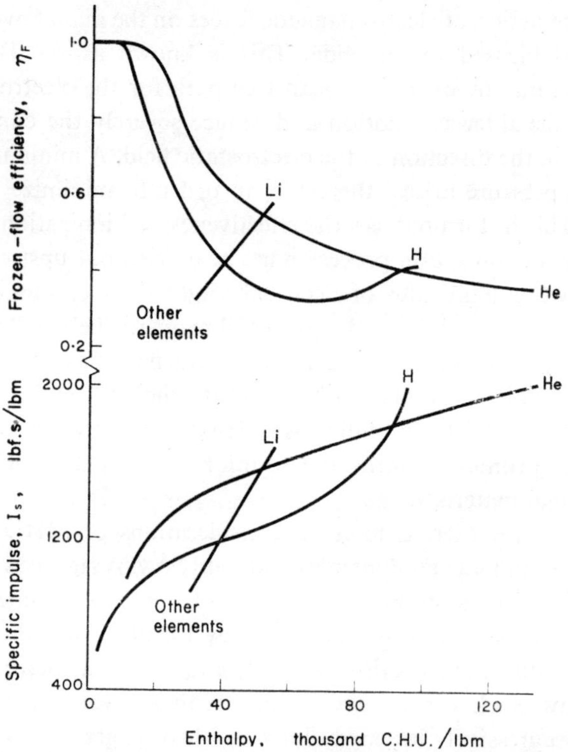

FIG. 5.13. Propellant performance in electrothermal rocket (derived from ref. 15).

energy, which can be provided thermally, permits the electrons to escape from the atomic structure. However, the temperature levels necessary are of order 2000° K, and this poses severe problems in connection with electrode and constructional materials. Ionization is assisted, therefore, by seeding the working fluid with vapour of alkali metals (e.g. caesium or potassium), and also by the appropriate selection of pairs of low ionization-energy propellant and high work-function surface, as shown in Fig. 5.4.

The electrothermal rocket derives its thrust by gas expansion, and therefore is subject to the laws applicable to all thermal rockets. Consequently, thrust tends to be directly proportional to propellant molecular weight, and specific impulse to be inversely proportional. Thus, low molecular weight leads to propellant economy. Analytical results of the performance of the light elements at one atmosphere operating pressure[15] show that, up to about 40,000 CHU/lbm (170×10^6 J/kg), added enthalpy, helium and hydrogen exhibit the highest specific impulse and highest frozen-flow efficiency (Fig. 5.13). Lithium is likely to be more promising at higher levels of performance.

When the propellant is accelerated electrically, the same requirements of high molecular weight for thrust, and low molecular weight for specific impulse, are generally found to apply. Substitution of ΔV from equation (5.3) in equation (5.6), for example, shows that, for a given specific impulse, $F \propto \boldsymbol{m}^2$, hence a propellant of high molecular weight is suited to thrust and, for this reason, mercury, and heavy particles such as colloids, powders or large organic molecules have been proposed. For a given input of electrical energy, however, the specific impulse equation (5.4) shows the need for a propellant of low molecular weight. A balance must be struck, therefore, to meet these conflicting requirements in relation to the mission as a whole.

Symbols

A = Electrode area
B = Magnetic flux density
c = Velocity of light
D = Flux density
E = Electric field intensity
E_i = Internal e.m.f.
F = Force on particle
H = Intensity of magnetic field
I_{sp} = Specific impulse in consistent units
i = Current
J = Current density
k = Proportionality constant
k_B = Boltzmann's constant
l = Distance

m = Mass of propellant
$\boldsymbol{m}$ = Mass of particle
q = Strength of charge
R = Aspect ratio of beam
$\boldsymbol{r}$ = Radius of particle
T = Temperature
t = Time
u = Velocity of source charge
V = Electrical potential
v = Velocity of test charge
v_e = Effective velocity of propellant through exit plane
v_r = Root-mean-square velocity of particle
W = Work done
ε = Specific dielectric constant of material
ε_0 = Dielectric constant of vacuum
$\varnothing$ = Work function
μ_0 = Permeability of vacuum
τ = Volume

Subscripts

s = Electrostatic
m = Magnetic

CHAPTER 6

ADDITIONAL PROPULSION CONCEPTS

ADDITIONAL proposals for spacecraft propulsion range from the reasonably possible to the highly unlikely. In general, the attractions lie in ultra high efflux velocity and/or the utilization of energy freely available within the space environment. Possible disadvantages include excessive mass of propulsion system and crippling reduction of payload. A few transportable high-velocity systems are examined briefly below, followed by some systems based upon environmental sources of energy.

6.1. Transportable Systems

6.1.1. Nuclear Particle Emitters

Since nuclear particles can be emitted energetically by radioactive decay, nuclear fission or nuclear fusion, it is conceivable that directed emission might be possible by some collimating device in order to provide thrust by reaction from a particulate jet. The concept is attractive since emission velocities are high; those of fast neutrons of 1 MeV kinetic energy, for example, giving a specific impulse of order 10^6 lbf.s/lbm.

When v_e approaches the velocity of light, c, a relativistic increase in the mass of the efflux becomes apparent to an observer on the vehicle, with a corresponding increase in thrust, i.e.

$$F = \left(\frac{\dot{m}}{\sqrt{1 - v_e^2/c^2}}\right) v_e ,$$

where m is the rest mass of the efflux. In this case, thrust increases rapidly with v_e, but the kinetic power of the efflux, P_e, increases with v_e^2, and the ratio of thrust to efflux power, F/P_e, falls.

Since the penetration of such particles as electrons and helium nuclei is very short, a thin film of reactive fuel would be necessary, backed by a reflector sheet, in the form of a propulsive sail. An additional problem is the space charge build-up in the case of continuous emission of charged particles, and the consequent loss of thrust. On current technology, therefore, no major developments appear to be feasible with this type of propulsive system.

6.1.2. *The Photon Rocket*

When v_e eventually reaches the level of c, as with the ejection of photons, no efflux mass, apparently, is involved. However, the generation of a photon within the engine, and its subsequent decay in space, involves the mass-energy transfer expressed by Einstein's equation, $E = mc^2$, hence,

$$F = \dot{m}c = \left(\frac{\dot{E}}{c^2}\right)c = \frac{\dot{E}}{c} = \frac{P_e}{c},$$

$$I_{sp} = F/\dot{m} = c = 9{\cdot}84\times10^8 \text{ ft/s},$$

and $$I_s = c/k = 3{\cdot}06\times10^7 \text{ lbf.s/lbm} = I_w, \quad \text{s.}$$

It is interesting to consider the thrust/efflux power ratio at this stage:

$$\begin{aligned} F/P_e &= 1/c = 0{\cdot}102\times10^{-8} \text{ s/ft}, \\ &= 0{\cdot}561\times10^{-6} \text{ lbf/hp}, \\ &= 3{\cdot}34\times10^{-6} \text{ N/kW}. \end{aligned}$$

This shows clearly the very low order of thrust obtainable, even with a massive power input, e.g. half a pound of thrust for one million horsepower.

Photons may be generated from a source of high energy such as a very hot surface or a nuclear reaction. Since the generated photon flux will radiate isotropically, generation must be followed by direction control in order to approach a collimated rearwards-directed jet.

Photons are emitted when planetary electrons change to an orbit of lower energy level closer to the nucleus, and these photons could conceivably be reflected by a dense mirror forming the walls of the chamber and propelling nozzle. Photons of higher energy are emitted as γ-rays from nuclear reactions, and those emitted in unusable directions might be eliminated by absorption in a lead shield.

Thrust may also be expressed in terms of the frequency of the electromagnetic radiation associated with the photon, e.g.

$$F = \dot{E}/c = (h \cdot f)/c,$$

where h = Planck's constant = $6{\cdot}625 \times 10^{-34}$ J-s, and f = frequency of radiation.

Here again, very large quantities of energy are seen to be required even for nominal levels of thrust. Furthermore, energy absorption by the mirrors and shielding would present severe problems of cooling.

A special variant of the photon jet is the laser (light amplification by stimulated emission of radiation) device. This produces a very intense, highly directional, beam of light which is not only monochromatic but is coherent, i.e. the radiations are in phase. This is achieved by stimulation of the photon-producing atoms by means of an external magnetic field operating at the selected frequency. Planetary electrons are excited into higher energy orbits and, on return to their normal equilibrium condition, emit photons which, by reflection within a resonant cavity, are made to generate further photons until the beam penetrates the end of the cavity. The intensity of a laser beam is a million times brighter than sunlight.

6.1.3. Nuclear Bomb Propulsion

A system offering both high thrust and high specific impulse, with thrust/weight ratio values in excess of 2, is the "brute force" method of exploding a series of nuclear fission bombs some distance from the vehicle. A fraction of the initial waves of electromagnetic energy is captured by an ablative pusher platform, imparting an impulse to the vehicle from both the absorbed energy and the rearward expulsion of vaporized platform material. This is followed by an impulse from the propellant energized within the bomb, together with some fragments of the bomb itself. Payloads of up to 13,000 tons have been conjectured, with specific impulses of over 25,000 lbf.s/lbm. Obvious disadvantages are the inefficiency of propellant utilization, which is a function of the ratio of the solid angle of radiation and propellant capture to 4π, and contamination of space by radioactive fission products.

6.2. Environmental Systems

6.2.1. The Solar Heater Rocket

Solar radiation is currently used on a small scale for a variety of heating purposes ranging from metallurgical refining in inert atmospheres to the cooking of analfresco meal. On the same principle, solar radiation might be concentrated by a parabolic reflector onto a heat exchanger comprising part of a propellant system. Hydrogen or ammonia propellant would collect its heat from the exchanger and then expand through a propelling nozzle in the conventional way. The level of energy available is then a function of the area and effectiveness of the parabolic reflector, and the level of the solar radiation at that point in space.

Alternatively, the solar energy may be converted directly to electrical energy by means of photoelectric cells, and utilized to provide thrust in a small electric propulsion unit. Collection areas of several thousands of square feet are likely to be necessary for a nominal thrust of, say, one pound.

6.2.2. The Solar Sail

The incidence of electromagnetic radiation upon a surface results in a radiation pressure, as evidenced by the radiometer instrument. At the distance of one astronomical unit from the Sun, the level of radiation pressure of sunlight is approximately 10^{-7} lbf/ft^2 and, with normal reflection, this pressure tends towards double this value. Allowing for losses, a nominal thrust of one pound will require a reflecting area of approximately 10 million square feet, represented by a circle of about one-quarter mile diameter. Since solar energy is directional, propulsion by solar reflection might be achieved by the solar sail device on the same principle as the handling of wind by a ship's sail. The great attractions of this system are the environmental source of both power and propellant, and the resulting specific impulse value of infinity. Development of lightweight plastic sheeting, suitably coated with reflecting metals, gives promise of practicability of such a system, although the operational life under space conditions has yet to be determined. The thrust/weight ratio should be about 10^{-4} which is comparable to those of the much more complicated photon rocket and electrostatic propulsion systems.

6.2.3. *Ramjet Propulsion in Planetary Atmospheres*

The air-breathing ramjet is already a practicable propulsion device in service for the high-speed propulsion of missiles within the Earth's atmosphere. It is conceivable that this unit might lend itself to operation in the atmospheres of other planets, containing such fuel materials as methane and ammonia, by the provision of a suitable vehicle-borne oxidant material with which a chemical reaction can take place. There are also possibilities for the recombination ramjet designed to operate within the narrow layer of free radicals surrounding the outer limits of the Earth's atmosphere.

6.2.4. *Gravitational Fields*

There is as yet no evidence to suggest the possibility of matter with an independence of gravitational forces, nor with a negative gravitational mass. Such a discovery would open the way to propulsion by very low-weight thrust devices in the former case, and by gravitation itself in the latter. Prospects of developments in this direction are slight, but there is still much to be learned about the subject of gravity. One possible application would be vehicle direction-changing by tangential flight through the gravitational field of a nearby astronomical body.

6.2.5. *Magnetic Fields*

Theoretically, the passage of electric current through conductors within a vehicle traversing a magnetic field could provide a propulsive thrust. However, the minute strength of interplanetary magnetic fields (approximately 10^{-9} Wb/m^2) gives little promise of any realistic performance.

Symbols

c = Velocity of light
E = Energy
F = Thrust
f = Frequency of radiation
h = Planck's constant
I_s = Specific impulse in units of lbf.s/lbm
I_{sp} = Specific impulse in consistent units
I_w = Specific impulse on weight flow basis
m = Mass of propellant
P_e = Kinetic power of efflux
v_e = Effective velocity of propellant through exit plane

CHAPTER 7

SUMMARY AND CONCLUSIONS

7.1. Propulsive Systems

The basis of most systems of spacecraft propulsion is the conversion of available energy to the kinetic energy of a propellant jet. Several conversion routes are possible, involving various intermediate stages. Those systems based upon transportable energy sources are represented broadly in Fig. 7.1, together with the names given to the types of rocket in which they are utilized.

In essence, the ultimate performance of the chemical rocket is limited by the energy content of the chemical reactants, and thus upon their bond energies. Specific impulse is taken as the most reliable and representative parameter of reactant performance, but some indication of performance is seen to be obtainable from the heating value expressed on the basis of unit mass of reactant and, to a lesser extent, on unit mass of fuel. The release of the far greater energy content of nuclear reactants, however, is not proportionate due to the thermal problems with structural materials, although these are partly overcome by gaseous fission in the absence of wall contact. Release of the additional energy available from nuclear reactants is most likely in the electrical form, and this can be applied either thermally, by direct application in a reaction chamber as in gaseous fission, or by means of electrostatic or magnetic acceleration using an electrically conducting plasma as propellant. In ordeı of presentation above, these systems show an increasing level of specific impulse.

Since efflux power bears some relation to available source power, and since $P_e = \frac{1}{2} F I_{sp}$ (Chapter 1), a given availability of power may be utilized in a variety of ways, with emphasis on either thrust or specific

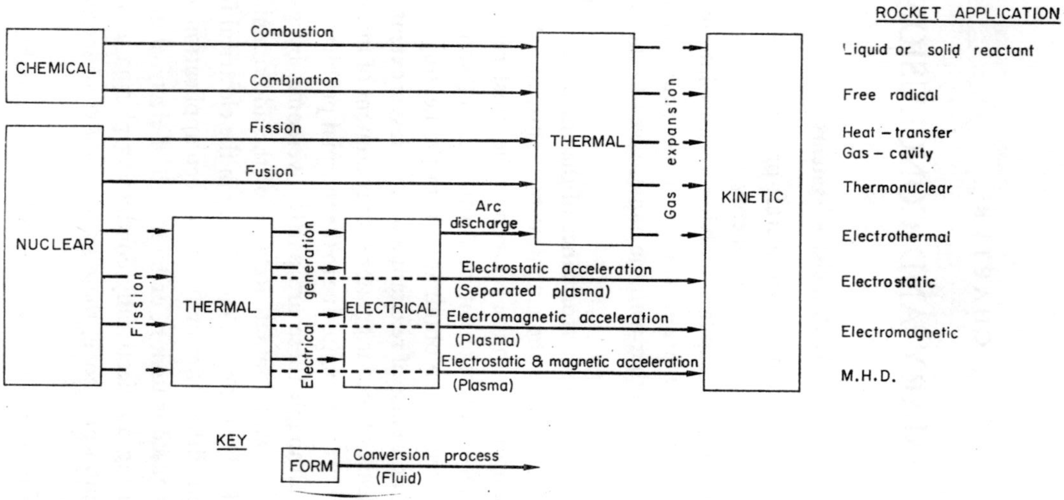

FIG. 7.1. Transportable energy forms and conversions for representative spaceflight propulsion systems.

TABLE 7.1. Propulsion System Performance (typical maximum values)[13]

System		Thrust, F (lbf)	Specific impulse, I_s (lbf. s/lbm)	Thrust/weight ratio, F/W	Typical propellants
Chemical	Liquid	2×10^6	410	100	Products of
	Solid	5×10^4	260	200	C, H, O, N & F
Nuclear	Heat transfer	10^6	1200	30	H_2
	Gas cavity		3000	10	U fission products
	Thermo nuclear		3000	0·1	Li
Nuclear electric	Thermal	10	2500	0·01	He, Li, H_2, NH_3
	Static	1	25,000	0·0001	Cs, K, Na
	Magnetic	1	15,000	0·001	
Solar	Heater	1	800	0·01	H_2
	Sail	1	∞	0·0001	Photons

impulse. Thus, a rectangular hyperbola can be imagined superimposed on Fig. 1.9. In general, therefore, the increasing level of specific impulse noted above tends to reflect a reducing level of thrust. Some representative values of specific impulse and thrust are listed in Table 7.1.

The general picture of energy conversion to electricity is presented in Fig. 7.2, showing the elimination of stages resulting from methods of direct conversion. The heavy power requirements for electrical propulsion are likely to be met initially by nuclear fission–turbine-induction, and later, perhaps, by direct solar photocell conversion.

The levels of energy *required* for spaceflight, expressed on the basis of unit mass of vehicle, were shown in Fig. 1.14 for representative missions. The levels of energy *available*, per unit mass of transportable reactant, have been derived in subsequent chapters, and are plotted on a compar-

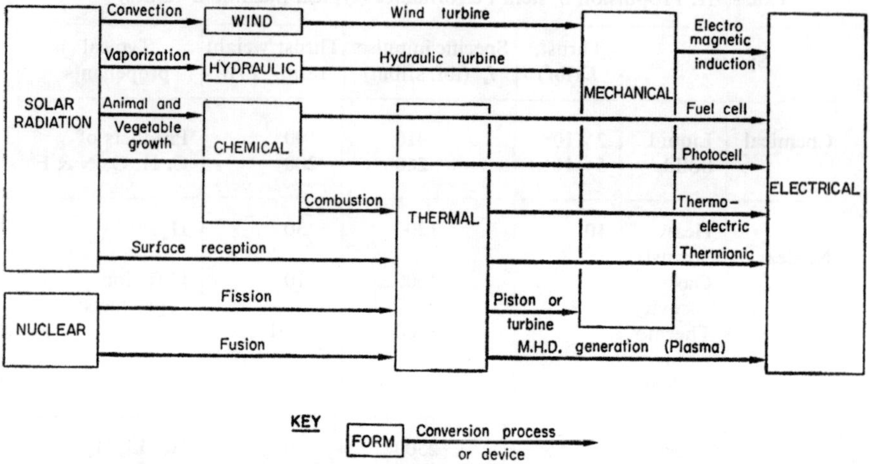

FIG. 7.2. Energy forms and conversions to electricity, showing elimination of stages by "direct conversion".

ative basis in Fig. 7.3. The task of the propulsion system, therefore, is to extract the maximum possible energy from each unit mass of reactant carried, and impart the maximum possible proportion of it to each unit mass of the parent vehicle. The relationship between these "specific energies" may be derived approximately in the following manner.

Let E_R = stored energy per unit mass of reactant, and
V_b = burnout velocity of vehicle.

$$\text{Then,}\quad \frac{\text{energy absorbed per unit mass of vehicle}}{\text{energy released per unit mass of reactant}} = V_b^2/2E_R = \varepsilon,$$

assuming no work is done against gravity (i.e. no change in potential energy).

But, $V_b = v_e \ln M_i/M_f$ (the rocket equation), for cases where v_e is constant.

$$\text{Thus,}\quad \varepsilon = \frac{(v_e \ln M_i/M_f)^2}{2E_R}.$$

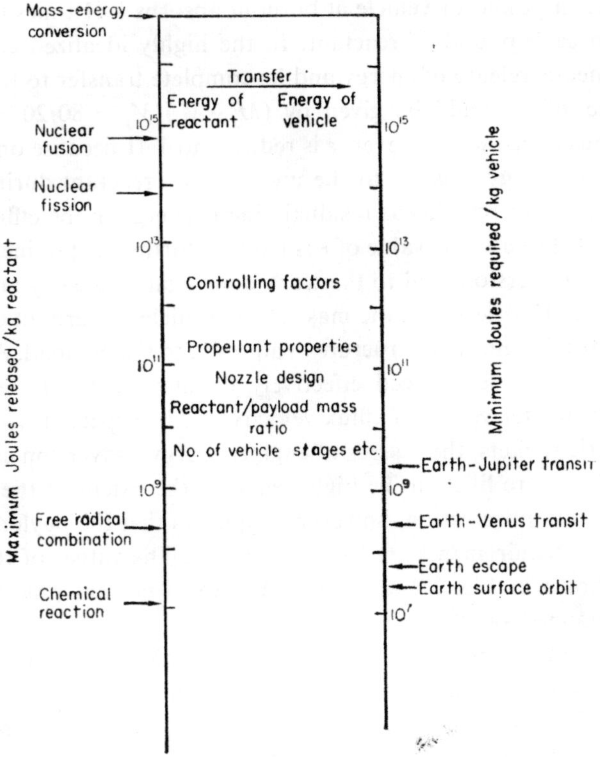

FIG. 7.3. Energy transfer from reactant to vehicle (ref. 13).

In chemical rocket practice, the mass of reactant forms a high proportion of the initial mass of the vehicle such that, in favourable circumstances, the value of ε is greater than unity. Using the performance figures for liquid hydrogen and liquid oxygen reactants from Chapter 3, for example, with a reactant mass proportion of 80 per cent of the initial launch mass, gives

$$\text{Peak} \quad I_s = 390 \text{ lbf.s/lbm} \equiv 12{,}500 \text{ ft/s},$$

$$E_R = 3190 \text{ CHU/lbm reactant mixture},$$

and

$$M_i/M_f = \frac{(12{,}500 \times 1{\cdot}610)^2}{32{\cdot}17 \times 1400 \times 2 \times 3190} = 1{\cdot}41.$$

Hence, each pound of vehicle at burnout absorbs 1·41 times the energy stored in each pound of reactant. In the highly idealized case of the instantaneous release of energy and its complete transfer to the vehicle, the value of ε would be given by $(M_f - M_i)/M_f = 80/20 = 4$. With continuous thrusting, however, ε is reduced to 1·41 because one portion of released energy is given to the unconsumed reactant during thrusting, and another remains as residual kinetic energy in the efflux.

In a realistic case, the value of ε is reduced further due to inefficiencies of energy conversion, and to the work done against gravity and atmospheric drag. Furthermore, the mass of the vehicle at burnout comprises the masses of engine and structure in addition to the payload. As shown, the former can be reduced effectively by means of vehicle staging. Although greater values of efflux velocity will be expected with nuclear and electric rockets, the masses of engine, energy conversion equipment and shielding are likely to be high, reducing the extent of the payload. With careful optimization, however, it appears that the payload masses and energy requirements are likely to be met. The values of thrust and weight shown in Fig. 7.4 are representative of current practice in the field of manned satellites.

The various manoeuvres comprising a complete mission have been shown to require individual optimization. Initially, a high thrust is necessary in order to approach the escape condition but, subsequently, the choice of variables for optimization is wide. For deep space missions, maximum payload forms the most significant requirement, and one of the major controlling variables is the velocity of the ejected propellant. This in turn is related back to the fraction of total mass available for the propulsion system, hence a gross increase in payload is restricted to a net increase by the additional mass of power plant needed to achieve it. A further important requirement, particularly with manned spaceflight, is the minimum duration of the mission, and such cases call for optimization of both payload and duration.

In general, vehicle propulsion systems are characterized by either short-duration high thrust giving high-impulse trajectories, or long-duration low thrust giving spiral orbits about the main gravitating body. The former represent lower overall energy requirements, but may well

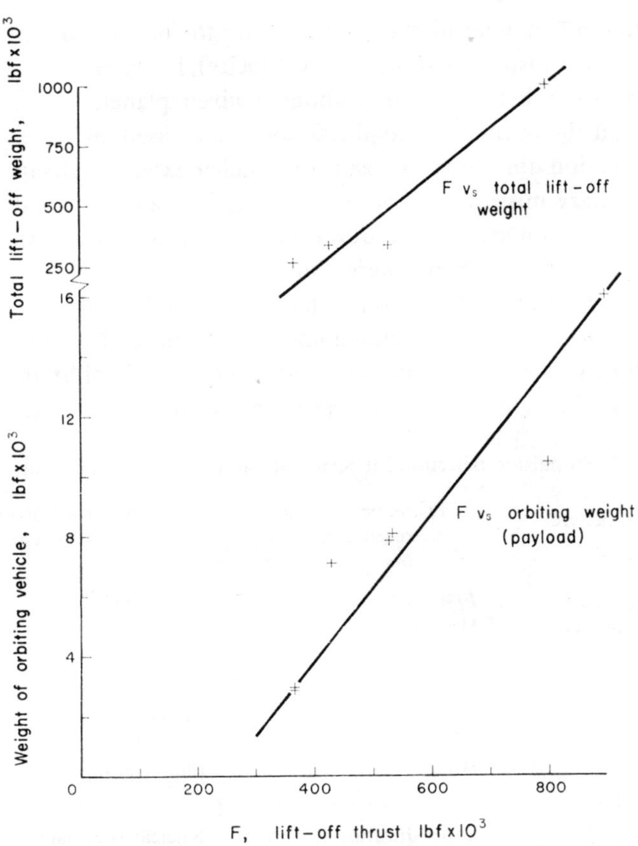

FIG. 7.4. Representative thrusts and weights for manned orbiting vehicles. (Derived from refs. 11, 26 and private communication from N.A.S.A., Washington, D.C.)

be restricted by upper limits to energy content per unit mass of reactant, to the power plant mass and hence the power output, or to the operating temperature of the thruster materials. Examples of these limitations are to be found in chemical, electric and nuclear-heat-transfer rockets respectively.

From a theoretical analysis of space vehicles with no such restrictions placed on exhaust velocity, Cruddace[27] shows that maximum payload results when the exhaust velocity is programmed to rise during thrust operation in a manner to maintain constant the product (instantaneous vehicle mass × instantaneous exhaust velocity), i.e. $v_e = (M_0/M)\, v_{e_0}$.

For transfer between orbits about a given planet, Cruddace also shows that the optimal payload fraction is increased, at the expense of longer mission duration, by means of a higher exhaust velocity. For an interplanetary mission, on the other hand, increased payload fraction results from a longer mission duration but at a lower exhaust velocity.

Propulsion requirements were shown very briefly in Table 1.3, and selection of suitable propulsion systems is shown broadly in Table 7.2, more detailed selection depending upon the purpose of the mission and on such factors as cost limitations and reactant availability in addition to payload fraction and required mission duration.

TABLE 7.2. Propulsion Selection for Space Missions with High Payload Fraction

Manoeuvre	Prime performance requirement	Appropriate propulsion system
First stage from launch planet to orbit	$F/W > 1$ High F	Liquid chemical
Orbital adjustment	Short duration ↓ High I_s	Electrothermal Colloid electrostatic or MHD Electrostatic
Launch planet orbit to satellite orbit	Short duration (manned)	Nuclear thermal
	High I_s (unmanned)	Electrostatic
Launch planet orbit to target planet orbit	High I_s	Electrostatic or solar sail
Orientation control	High I_s	Electrostatic

(Derived from ref. 27.)

7.2 Outlook

The discussion outlined in this book crystallizes into the realization that the energy requirements for spaceflight *can* be met by the energy sources currently available, and the processes of conversion to thrust which already appear practicable. Spaceflight, within the solar system at least, is therefore certain, since the human race has invariably met the technological challenges brought within its attainment. Critics may well question the expense of such an operation, but recent history already shows an abundance of benefits that can be expected to flow from an enterprise of such magnitude and inspiration; not only obvious benefits of a scientific, technological and biological nature, but the more abstract improvements in political and national relationships which could evolve from the co-operative effort necessary between different countries. Transportation throughout space could be man's greatest and noblest achievement, and today's population is privileged to be in at the curtain raising.

Very many problems remain to be solved, not the least being the continuing provision of adequate funds in the light of overall human needs, but the momentum towards eventual success is clearly evident. The vast countdown extending over the centuries is steadily approaching zero. Without any doubt, spaceflight will very shortly be GO.

Symbols

E_R = Stored energy per unit mass of reactant
I_s = Specific impulse in units of lbf.s/lbm
M_i = Initial mass of vehicle
M_f = Final mass of vehicle
V_b = Burnout velocity of vehicle
v_e = Effective velocity of propellant through exit plane
ε = Specific energy ratio

APPENDIX

THE TREATMENT OF VECTOR QUANTITIES

WHEREAS a scalar quantity can be described completely by its magnitude and algebraic sign, a vector quantity requires the specification of direction in addition. A vector is commonly expressed in bold-face type and represented diagrammatically by an arrow of appropriate length and direction. A vector can be located precisely in space by reference to a coordinate framework represented, say, by x, y and z. A velocity vector, **V**, for example, would have three mutually perpendicular components of magnitude V_x, V_y and V_z and direction i_x, i_y and i_z, respectively, where i_x, etc., being unit vectors, indicate direction but do not influence magnitude. Hence, the velocity vector **V** is completely defined as follows;

$$\mathbf{V} = V_x i_x + V_y i_y + V_z i_z .$$

Addition of Vectors

Each vector of a group of vectors will have components parallel to the three coordinate axes. Hence, addition of the vectors becomes a sim-

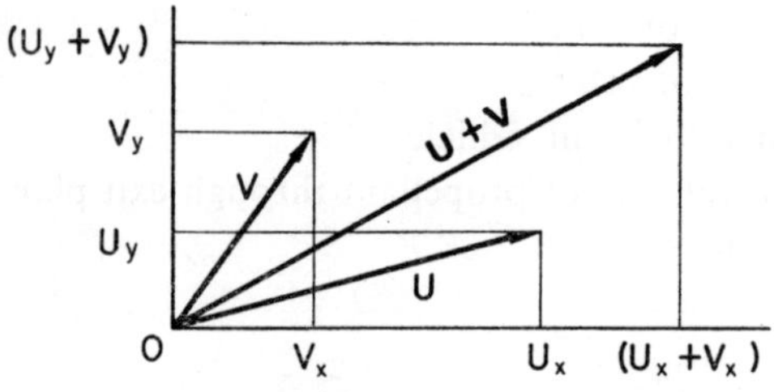

FIG. A.1. $\mathbf{U}+\mathbf{V} = (U_x + V_x)\, i_x + (U_y + V_y)\, i_y$.

ple summation of the components, e.g.

$$\mathbf{U}+\mathbf{V} = (U_x+V_x)\,i_x+(U_y+V_y)\,i_y+(U_z+V_z)\,i_z.$$

The two-dimensional case is shown in Fig. A.1.

Multiplication of Vectors

The two important interpretations of vector multiplication are the dot product, written for the above vectors as $\mathbf{U}\cdot\mathbf{V}$, and the cross product written $\mathbf{U}\times\mathbf{V}$. The former is a scalar quantity and comprises the sum of the scalar products of the corresponding components, e.g.

$$\mathbf{U}\cdot\mathbf{V} = U_xV_x+U_yV_y+U_zV_z.$$

The two-dimensional case is shown in Fig. A.2.

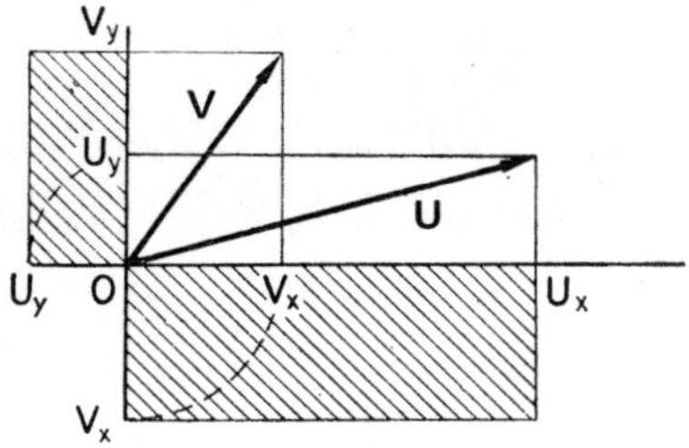

FIG. A.2. $\mathbf{U}\cdot\mathbf{V} = (U_xV_x+U_yV_y)$.

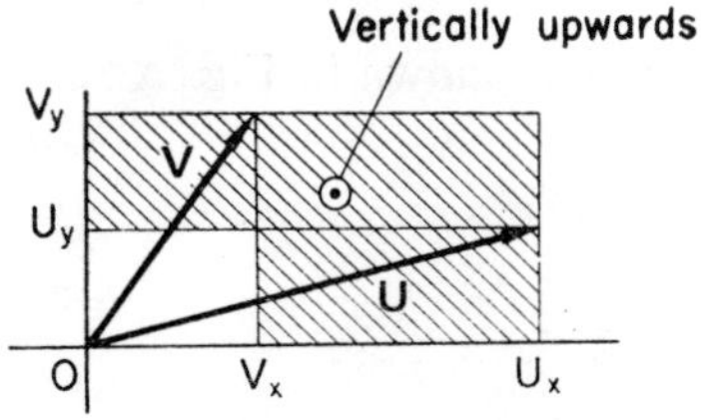

FIG. A.3. $\mathbf{U}\times\mathbf{V} = (U_xV_y-U_yV_x)\,i_z$.

The cross product is a vector quantity of magnitude equal to the sum of the rectangular areas formed by the components, each area being directed perpendicularly to its plane in the same sense as a right-hand thread, i.e. the cross product $(i_x \times i_y) = i_z = -(i_y \times i_x)$ for the coordinate axes in Fig. A.4. Consequently, $(i_x \times i_x)$, $(i_y \times i_y)$, etc., are zero since they encompass no area.

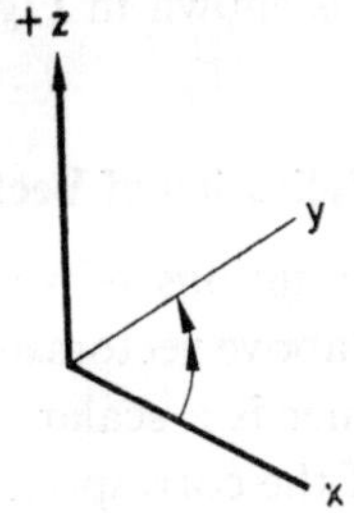

FIG. A.4. Right-hand rule.

For the general case of vectors having components on all three co ordinate axes, the cross product becomes:

$$\begin{aligned}\mathbf{U}\times\mathbf{V} &= (U_x i_x + U_y i_y + U_z i_z)\times(V_x i_x + V_y i_y + V_z i_z)\\ &= (U_x V_x)(i_x\times i_x) + (U_x V_y)(i_x\times i_y) + (U_x V_z)(i_x\times i_z)\\ &\quad + (U_y V_x)(i_y\times i_x) + (U_y V_y)(i_y\times i_y) + (U_y V_z)(i_y\times i_z)\\ &\quad + (U_z V_x)(i_z\times i_x) + (U_z V_y)(i_z\times i_y) + (U_z V_z)(i_z\times i_z).\end{aligned}$$

Using the previously derived values for $(i_x \times i_x)$, $(i_x \times i_y)$, etc.,

$$\mathbf{U}\times\mathbf{V} = (U_y V_z - U_z V_y)\, i_x + (U_z V_x - U_x V_z)\, i_y + (U_x V_y - U_y V_x)\, i_z .$$

The two-dimensional case is shown in **Fig.** A.3.

REFERENCES AND BIBLIOGRAPHY

1. ANON., *Inter-Avia*, Vol. XI, No. 6 (1956).
2. ANON., *International Critical Tables*, Vol. V, p. 163, New York: McGraw Hill.
3. ANON., *JANAF Thermochemical Tables*, Dow Chemical Co., Michigan, U.S. (1965).
4. ANON., Rocketdyne Co., U.S.A., Theoretical performance of rocket propellant combinations. Chart published 1 Jan. 1959.
5. ANON., *U.S. Standard Atmosphere*, Govt. Printing Office, Washington, D.C. (1962).
6. BOSTICK, W. H., Experimental study of plasmoids, *Phys. Rev.* 106, 404 (1957).
7. BROWN, G. I., *A Simple Guide to Modern Valency Theory*, Longmans Green & Co. (1957).
8. BUSSARD, R. W. and DE LAUER, R. D., *Nuclear Rocket Propulsion*, McGraw-Hill Book Co. Inc. (1958).
9. CAHEN, G. and TREILLE, P., *Nuclear Engineering*, Allyn & Bacon, Inc. (1961).
10. CORLISS, W. R., *Propulsion Systems for Space Flight*, McGraw-Hill Book Co. Inc. (1960).
11. GATLAND, K. W., *Spacecraft and Boosters*, Iliffe (1961, and 2nd volume 1965).
12. GOODGER, E. M., Property requirements for liquid rocket propellants, *J. Inst. Pet.* **46,** No. 442 (Oct. 1960).
13. GOODGER, E. M., High-energy propulsion for spaceflight, *1966 Fuel Conference, Institute of Fuel, Newcastle, N.S.W.* (Oct. 1966).
14. HAVILAND, R. P. and HOUSE, C. M., *Handbook of Satellites and Space Vehicles*, D. van Nostrand Co. Inc. (1965).
15. JACK, J. R., Theoretical performance of propellants suitable for electrothermal jet engines, *N.A.S.A. TN D-682* (1961).
16. MANSFIELD, W. K., *Elementary Nuclear Physics*, Temple Press (1958).
17. OFTERMATT, W. F., Tables of bond and resonance energies for estimating standard heats of formation, *Cal. Inst. Technology, AD-24936* (1953).
18. PEDERSEN, E. S., *Nuclear Energy in Space*, Prentice-Hall Inc. (1964).
19. PERCHONOK, E., Performance evaluation of ramjet propellants, *Chemistry of Propellants Meeting*, AGARD, Paris (June 1959).
20. ROTH, J. R., A preliminary study of thermonuclear rocket propulsion, *J.B.I.S.*, **18,** 99–108 (1961–2).
21. ROTHROCK, A. M., Spacecraft propulsion (contained in *Advanced Propulsion Techniques*, Ed. PENNER, S. S., Pergamon Press, 1961).
22. SHEPHERD, L. R. and CLEAVER, A. V., The atomic rocket, *J.B.I.S.*, p. 61 (1949).
23. SPROULL, R. L., *Modern Physics*, 2nd edn., John Wiley & Sons Inc. (1956).
24. SUTTON, G. P., *Rocket Propulsion Elements*, Wiley (1956).

25. TURCOTTE, D. L., *Space Propulsion*, Blaisdell Publishing Co. (1965).
26. ANON., *Astronautics and Aeronautics* (Feb. 1966).
27. CRUDDACE, R. G., The application of advanced propulsion systems to deep-space vehicles, *J.B.I.S.* **19** 133 **47** (1963–4).

Bibliography

ALPERIN, M. and SUTTON, G. P. (Eds.), *Advanced Propulsion Systems*, Pergamon Press (1959).

BRAGG, S. L., *Rocket Engines*, George Newnes Ltd. (1962).

DABOO, J. E., *Solid Fuel Rocket Propulsion*, Temple Press (1962).

HUMPHRIES, J., *Rockets and Guided Missiles*, Ernest Benn Ltd. (1956).

HUNTER, M. W., *Thrust into Space*, Holt, Rinehart & Winston Inc., N. Y. (1966).

N.A.S.A. SP-19, Chemical rocket propulsion, National Aeronautics & Space Administration, Washington, D. C., Dec. 1962.

N.A.S.A. SP-20, Nuclear rocket propulsion, National Aeronautics & Space Administration, Dec. 1962.

N.A.S.A. SP-22, Electric propulsion for spacecraft, National Aeronautics & Space Administration, Dec. 1962.

VALENTINE, J. M., *Atomic Physics*, English Universities Press (1960).

GLOSSARY

Activation energy. The energy required to initiate a reaction.

Activity. The rate of decay of radioactive nuclei.

Albedo. The fraction of incident solar energy reflected by a celestial body.

Alpha particle. The nucleus of the helium atom, He_2^4.

Antigravity. The concept of reducing the force of gravity.

Apogee. The point, on an elliptical orbit around a gravitating body, which lies furthest from the body.

Atomic Number (*Z*). The number of protons in a nucleus which, in a neutral electrically-balanced atom, is equal to the number of orbiting electrons.

Atomization heat. The energy required to divide an elemental molecule into its gaseous constituent atoms.

Ballistics. The movement of bodies propelled by impulsive thrust, and also the mechanism of propulsion by solid reactants.

Beta particle. An electron emitted by a nuclear reaction.

Binding energy. The energy emitted during the synthesis of a nucleus, represented by the loss of mass during the process.

Black body. A body which absorbs all incident radiation energy without reflection or transmission.

Bond energy. The energy required to divide a compound molecule into its gaseous constituent atoms.

Bremsstrahlung. The "braking radiation" created when charged particles are decelerated by electrostatic fields around heavy ions.

Burnout (or *Cutoff*). The conclusion of a period of thrust operation.

Characteristic exit velocity. The efflux velocity from a convergent–divergent nozzle equivalent to the fictitious thrust resulting when the full chamber pressure is assumed to apply at an equal throat area in a convergent nozzle. (N.B. In some references, the term "Characteristic Velocity" is used for the velocity of the vehicle.)

Choked nozzle. An expansion nozzle in which the gas velocity has reached sonic level at the throat.

Contact potential. The system of ion production by means of heated surfaces of high work function which remove the outermost electrons of the surrounding atoms of low ionization potential and release the resulting ions.

Critical mass. The minimum mass of material required for nuclear fission.

Cross-section, nuclear. The effective target area of the nucleus presented for reaction.

Cryogenic. The property of boiling and freezing at temperatures well below ambient.

Decay, radioactive. The spontaneous disintegration of a nucleus by the progressive emission of particles.

Delayed neutron. Neutron released by radioactive decay.

Deuterium. The atom of heavy hydrogen in which the nucleus incorporates a neutron in addition to a proton.

Deuteron. The nucleus of deuterium.

Dissociation. The decomposition of electrically-neutral molecules into component atoms and/or radicals on the addition of sufficient bond energy.

Double-base reactant. A solid reactant comprising a colloidal gel of two nitric esters.

Effective exit velocity. The efflux velocity from a thermal rocket with an ideal nozzle exit area to give an exit pressure equal to ambient.

Electron. A particle of negative electrostatic charge which orbits about the nucleus.

Electron bombardment. The system of ionizing a propellant flow by bombardment with high-energy free electrons.

Environmental engine. A reaction engine deriving any reactants or propellants from its environment.

Erosive burning. The combustion of a solid reactant surface under the scouring influence of the velocity of passing burnt gases.

Escape velocity. The velocity required by a vehicle to provide sufficient energy to overcome a gravitational field.

Expansion ratio. The ratio of exit area to throat area of an expansion nozzle.

Fast neutron. A neutron with an energy level above 10 keV.

Fission. The separation of a nucleus into two nuclei and a number of neutrons with the release of energy.

Free radical. Electrically-neutral atom or group of atoms with a high chemical reactivity.

Frozen (or *Fixed*) *equilibrium.* The condition where the concentrations of components of the propellant stream remain constant at their initial entry levels of chemical equilibrium throughout the nozzle flow.

Fusion. The combination of two nuclei into a single nucleus with the release of energy.

Galaxy. A large cluster of many stars and planets.

Gamma ray. Electromagnetic radiation of wavelength approximately 10^{-10} to 10^{-13} m produced by nuclear reactions.

Grain. A single shaped charge of solid reactant designed to fit into a rocket combustion chamber.

Half-life. The time required for the activity of a radioactive material to reduce to half of its initial value.

Hall effect. The mutually perpendicular drift imparted to charges crossing a magnetic field.

Hohmann transfer ellipse. The path of minimum propulsive energy between two planetary orbits.

Hypergolic. The property of spontaneous ignition when two reactants are brought into contact.

Ionization. The process of electron addition to, or removal from, an initially electrically-neutral atom.

Isobars. The nuclides of different elements which have a common number of nucleons.

Isotones. The nuclides of different elements which have a common number of neutrons.

Isotopes. The alternative forms of nuclide of the same element which have a common number of protons.

Magnetic mirror. A system of magnetic fields designed to contain plasma.

Magnetohydrodynamics. The combined science of electromagnetic and gas dynamic relationships.

Mass Number (A). The number of nucleons (protons plus neutrons) in a nucleus.

Mass ratio. The ratio of the initial mass of a vehicle or stage to the final mass after consumption of the reactants.

Meson. A short-lived particle within the nucleus involved in the production of a nucleic attraction force between the nucleons.

Moderator. Material which reduces the velocity of high-energy neutrons.

Momentum thrust. The component of thrust provided by the change in momentum of the propellant.

Neutron. A nucleic particle with no electrostatic charge.

Nucleon. A neutron or a proton.

Nuclide. A nuclear species capable of more than transient existence.

Obturator. A semi-seal device to permit equalization of pressure but prevent the flow of gas.

Orbit. The repetitive path taken in space by one body about another.

Overexpansion. The condition where the nozzle exit area is excessive such that the propellant pressure at exit tends to fall below ambient resulting in the formation of shock waves.

Payload fraction. The ratio of the mass of useful payload of a rocket vehicle to the total launch mass of the vehicle.

Perigee. The point, on an elliptical orbit around a gravitating body, which lies nearest to the body.

Photon. A quantum of electromagnetic radiation.

Planet. A major celestial body orbiting a star.

Plasma. A collection of high-energy atoms containing a relatively high proportion of free electrons and ions.

Platonization. The process of adjusting the components of a solid reactant to provide a relatively small variation in burning rate over a given range of pressure.

Potential energy well. The schematic representation of the variation of gravitational energy of two particles with the distance between them.

Pressure thrust. The component of thrust provided by the difference in the forward-acting static pressure of the propellant at the nozzle exit plane, and the reverse-acting atmospheric pressure over the same projected area at the front of the vehicle.

Propellant. The working fluid expelled from a propulsive system to provide thrust.

Proton. A nucleic particle with a positive electrostatic charge.

Radioactivity. The spontaneous disintegration of a nuclide.

Reactant. The material used to provide energy from its reaction and also, in the case of a chemical rocket, a source of propellants from its reaction products.

Recombination. The reassociation of atoms into electrically-neutral molecules following reaction and dissociation.

Relativistic effects. The increase in mass and the apparent slowing down of time observed on the reference frame moving with the particle at a velocity approaching that of light.

Resonance. A refinement of molecular structural theory which allows for complex atomic configuration not consistent with the simple concept of integral bonds.

Resonance capture. The relatively strong tendency to capture incident particles by an inelastic nuclear collision.

Rocket. A propulsion system in which thrust is obtained from the expulsion of propellant.

Satellite. A body moving in an orbit around a celestial body.

Shifting equilibrium. The condition where the concentrations of components of the propellant stream change to give continuous chemical equilibrium throughout the nozzle flow.

Specific impulse. The propellant economy of a rocket engine, expressed in terms of thrust per unit mass consumption rate of propellant.

Sputtering. Surface damage resulting from impingement of ions.

Stage. A single self-contained propulsion unit which, after consumption of its propellants, may be jettisoned from the main vehicle.

Star. A self-luminous body of gas in space.

Thermal neutron. A neutron with an energy level below 1eV which is in thermal equilibrium with its surroundings.

Thrust. The propulsive reaction force produced by the expulsion of propellant.

Thrust coefficient. The ratio of the actual thrust to the fictitious thrust resulting from the product of the chamber pressure and the area of the nozzle throat.

Total impulse. The integral of thrust with respect to time.

Underexpansion. The condition where the nozzle exit area is insufficient such that the propellant pressure at exit exceeds ambient leading to the formation of external expansion waves.

Volume (or *Density*) *impulse.* Specific impulse expressed in terms of volume consumption rate of propellants.

Weight impulse. Specific impulse expressed in terms of the weight consumption rate of propellants referred to sea-level conditions at Earth.

Work function (or *Surface work*). The electric potential equivalent to the minimum work required to free the most energetic electron from the surface of a solid material.

INDEX

OTHER TITLES IN THE SERIES

DIV. III. PROPULSION SYSTEMS INCLUDING FUELS

Vol. 1. Penner — *Chemistry Problems in Jet Propulsion*

Vol. 2. Ducarme & Lefebvre — *Progress in Combustion Science and Technology*

Vol. 3. Ragozin — *Jet Propulsion Fuels*

Vol. 4. Kholshchevnikov — *Some Problems in the Theory and Assessment of Turbo-Jet Engines*

Vol. 5. Zuyev & Skubachevskii — *Combustion Chambers for Jet Propulsion Engines*

